Enzyme Immunodiagnosis

Enzyme Immunodiagnosis

EDOUARD KURSTAK

Comparative Virology Research Group
Faculty of Medicine
University of Montreal
Montreal, Quebec
Canada

1986

ACADEMIC PRESS, INC.

Harcourt Brace Jovanovich, Publishers

Orlando San Diego New York Austin
Boston London Sydney Tokyo Toronto

ACADEMIC PRESS, INC.
Orlando, Florida 32887

United Kingdom Edition published by
ACADEMIC PRESS INC. (LONDON) LTD.
24–28 Oval Road, London NW1 7DX

Library of Congress Cataloging in Publication Data

Kurstak, Edouard.
Enzyme immunodiagnosis.

Bibliography: p.
Includes index.
1. Immunoenzyme technique. 2. Immunodiagnosis.
I. Title. [DNLM: 1. Immunoenzyme Technics.
QY 250 K96 e]
QP519.9.I44K87 1986 616.07'56 85-28805
ISBN 0–12–429745–5 (alk. paper)

PRINTED IN THE UNITED STATES OF AMERICA

86 87 88 89 9 8 7 6 5 4 3 2 1

Contents

5 Enzyme Immunohistochemical Methods

6 Enzyme Immunoassays after Immunoblotting

7 The Use of Polyclonal or Monoclonal Antibodies in Enzyme Immunoassays

8 Homogeneous Enzyme Immunoassays for the Detection of Drugs

Preface

Enzyme immunoassays are rapidly increasing in popularity and are becoming indispensable in many disciplines. They allow the measurement, with high sensitivity and specificity, of antibodies and antigens that may comprise an unlimited range of molecules. Moreover, enzyme immunoassays (EIA) provide a link between serological techniques and histological methods at both the light microscopic and the electron microscopic levels, and they are finding new applications in biochemistry and molecular biology. The development of monoclonal antibody production techniques has made the use of enzyme immunoassays even more advantageous and has accelerated their applicability. Not surprisingly, these assays have led to appreciable advances in many areas.

Unfortunately, the literature on this subject is almost completely restricted to primary sources, and overviews are lacking. This review attempts to fill this long-standing need by providing details of the basic technology needed to execute these assays and an overview of the applicability of enzyme immunodiagnosis to various fields. It is impossible to be exhaustive because of the applicability of EIA to almost all biological disciplines.

Due to the current great interest in their applicability to medical diagnosis and paramedical research, this book stresses research in these disciplines. Special attention is given to the applications of enzyme immunodiagnosis procedures to the detection of drugs, infectious disease pathogens, and antigens in cancer pathology. The use of immunoassays in vaccine assessment and standardization is also discussed, as are the commercial EIA kits, the automation of assays, and the interpretation of results obtained with both serological and immunohistochemical methods. The new developments in immunoblotting applications to diagnosis and research are also reviewed. The use of polyclonal or monoclonal antibodies in enzyme immunodiagnosis is discussed in depth. We hope this book will contribute to the spread of this very powerful immunoenzymatic technology to other scientific endeavors.

Those involved in research, development, teaching, and diagnosis in fields in which immunoassays are already in use and in fields in which immunoassays can be applied to great advantage will find this book of interest.

Drs. Christine Kurstak and Peter Tijssen participated in my research and helped in the collation of data and in the preparation of this manuscript. Both learned enzyme immunoassay techniques in my laboratory and subsequently aided me greatly in the clinical application and fundamental research of EIA procedures. I would like to express my thanks and sincere gratitude to them for several years of fruitful collaboration, resulting in numerous publications.

My research, presented in this volume, was supported by grants from the Natural Sciences and Engineering Research Council of Canada, the National Cancer Institute of Canada, and the Medical Research Council of Canada. Sincere thanks are extended to these institutions.

I also wish to express my sincere gratitude to the staff of Academic Press for their part in the production of this volume.

Edouard Kurstak

Enzyme Immunodiagnosis

1

Introduction

Enzyme immunodiagnosis is expanding very rapidly. It has been estimated that commercial kits, in which monoclonal antibodies are used, will represent in 1987 a market value of 500 million dollars in the United States and 2 billion dollars in 1990 [Sasson (Unesco), 1983]. Though enzyme immunodiagnosis is relatively simple and very sensitive, potential pitfalls for the unaware are numerous.

Enzyme immunoassays were introduced about 20 years ago. The principles on which they are based were initially quite similar to those of immunofluorescence, i.e., an antigen is specifically detected due to the discriminatory power of antibodies, and a marker attached to these antibodies indicates that such a reaction takes place. Enzymes are used as markers in enzyme immunoassays since they have, in contrast to fluorescent labels, the capacity of amplification. Since peroxidase was used as the enzyme, these techniques were known as immunoperoxidase procedures (Kurstak *et al.*, 1969). Other than the use of a different marker, this approach did not differ essentially from immunofluorescence.

An important advance was the use of various solid phases (originally plastics such as polystyrene) that have the capacity of binding antigens or antibodies in high concentrations in certain ionic conditions. These immunosorbents could then extract the molecule to be detected from the test fluid (in conditions where these sample molecules would not attach nonspecifically to the solid phase), which, in turn, could absorb an enzyme-labeled antibody (detector molecule). Instead of antigens in cells, the simple means now available to adsorb antibodies to the solid phase could be used to apply the immunoperoxidase method to this solid phase. This approach has many advantages: (i) simple immobilization instead of the time-consuming and expensive method of growing antigens in cells; (ii) antigens which cannot be grown in cells, as well as antibodies, can be immobilized; (iii) pure antigens can be immobilized, i.e., without interference from cross-reacting molecules as sometimes is encountered in cells, to high concentrations; (iv) quantitative analysis on the solid phase is feasible; (v) automation is possible.

A second important advance was the use of the enzyme as an antigen in the assay. A properly chosen incubation sequence of the various reagents makes conjugation of the marker to antibody superfluous (nevertheless, chemical conjugation remains popular). The antigenicity of the enzyme can also be used in several ways to increase the sensitivity and specificity of the assay.

A third important advance (Rubenstein *et al.*, 1972) was the development of "homogeneous" enzyme immunoassays which do not necessitate the use of a solid phase. These assays are based on antibody-mediated changes in enzyme activity. For example, a small antigen (e.g., a drug) is conjugated to lysozyme. This conjugate is enzymatically active (degradation of bacterial cell walls) in the absence of antibody to this drug. The addition of antibody (directed to the hapten) to the conjugate will inhibit the enzyme's ability to hydrolyze the large substrate. The antibody is prevented from inhibiting the enzyme if there is a competing antigen (hapten) in the sample. The presence of antigen is thus directly related to enzyme activity.

The development of the procedures to produce monoclonal antibodies has had a greater impact on improving the formulation of enzyme immunoassays than on most other serological procedures. Auxiliary recognition molecules (avidin/biotin, protein A, lectins) also contribute to the rapidly expanding use of enzyme immunoassays.

Some of the designs developed have in turn greatly contributed to our understanding of the function of the various immunoglobulins. For example, the paradigm that IgM-antibodies are indicative of a recent infection and IgG of a recurrent infection appears to be decreasingly justified (Meurman, 1983).

The detectability, sensitivity, specificity, reproducibility, precision, and detectability indices may depend on many, often interrelated, factors. In general, competitive assays tend to increase specificity and to decrease sensitivity, whereas the contrary holds for the noncompetitive assays.

The enzymes chosen for enzyme immunoassays differ from laboratory to laboratory. As a general rule, peroxidase is most sensitive for colorimetric assays, whereas β-galactosidase is most sensitive for fluorimetric assays. However, other factors are important in the choice. For example, for the detection of antigens in plant extracts, peroxidase is generally not advisable since high background levels (endogenous peroxidase) can be expected. Alkaline phosphatase is, therefore, often used in plant virology.

In this monograph, applications of enzyme immunoassays are discussed. An attempt is made to avoid enumeration of applications, to present approaches used in the field of enzyme immunoassays, and to demonstrate the applicability of these very powerful techniques.

Selected procedures for the preparation of reagents and the principles for the design of the various assays for serology and immunohistochemistry are first given in detail. Discussion of the applications is restricted to those representative of alternatives in design or approach; others are given in tables. For some of the applications a representative flow chart is given.

It is interesting but not surprising that a particular type of enzyme immunoassay has found greater application in one field than in another, depending on techniques used before enzyme immunoassays were available. For example, in pathology, enzyme immunoassays are used in conjunction with cytochemical procedures, and immunoperoxidase techniques became very important in this area. On the other hand, in virology serological procedures were popular but became rapidly replaced by enzyme immunoassays such as ELISA (Kurstak *et al.*, 1986).

Attention is drawn to the use of monoclonal antibodies with respect to their potential as well as to possible problems they present. The procedures for the production and characterization of monoclonal antibodies are, however, not within the scope of this monograph. However, the particular properties of monoclonal antibodies pertaining to their application in enzyme immunoassays are stressed. The impact of monoclonal antibodies is presently greatest with complex antigens (e.g., infectious agents) or complex systems (immunohistochemistry), whereas assays for simple antigens (haptens, drugs) are still least affected.

Immunohistochemical procedures in pathology have proved particularly valuable for the diagnosis of renal and skin diseases, the detection of specific tumor antigens, and the detection of anti-tissue antibodies in the sera of patients. These methods enabled the specific recognition of antigens, thus offering evident advantages over the traditional nonspecific cytochemical staining procedures.

In contrast, the advantage of enzyme immunoassays which contributed most to their application for the detection of drugs is the nature of the marker, i.e., homogeneous, competitive enzyme immunoassays are replacing radioimmunoassays due to the expense and hazards of the latter. Homogeneous, competitive enzyme immunoassays are intrinsically more specific than heterogeneous noncompetitive assays, albeit up to about 10^4 times less sensitive. The application of homogeneous competitive assays is in large part due to the specificity and sensitivity requirements of drug tests. Sensitivity generally is not a problem at the concentrations of a drug in the therapeutic range, but specificity for a given drug is very important. Particular problems may be posed by the metabolic conversion of these drugs.

Enzyme immunodiagnosis is increasingly being applied in the field of infectious diseases. These methods are particularly prominent in virology and parasitology, but still less in bacteriology. In bacteriology, problems with cross-reactivity of the sera, on the one hand, and the ease of many direct identification methods, on the other, have delayed the application of enzyme immunoassays. The advent of monoclonal antibodies could be particularly beneficial in this area.

A new important area in which enzyme immunoassays will be applied is in molecular biology or biotechnology. Highly sensitive probes can be developed to detect, e.g., a single copy of a particular DNA sequence per cell. These dotting procedures could replace the radioactive probes in conventional hybridization methods. Moreover, detection methods are being developed to detect clones containing recombinant DNA.

Preparation of Immunoglobulins, Antibodies, or Fab Fragments

Immunoglobulins account for about 10% of the serum proteins, of which, after immunization, 1–25% are antibodies. Unfortunately, the immunoglobulins are very heterogeneous, complicating the protocols for the purification of antibody. Several methods are, nevertheless, suitable for purifying immunoglobulins and thus antibodies. The bulk of serum proteins are acidic in contrast to the immunoglobulins, making ion-exchange chromatography an obvious step for purification procedures. The procedures required for the purification of monoclonal antibodies may be quite different from those employed for the polyclonal antisera since the latter represent an average of the properties of the individual antibodies. Standard procedures will, therefore, almost always be applicable for polyclonal antibodies but not necessarily for monoclonal antibodies.

A. IMMUNIZATION

The purpose of immunization is to obtain high-titered antisera with a high avidity. The properties of the antiserum are largely determined by the genetic composition of the animals (particularly the *Ir* genes). Inbred mice are good responders to certain immunogens, but respond badly to others. This is equally true for other animals, as shown by Harboe and Ingild

(1983) for outbred rabbits, and for man (Gonwa *et al.*, 1983). Up to a certain degree, an increase in the immunogen dose will also increase the antibody titer. Above this level, greater doses will increase cross-reactivity or other nonspecific activity without increasing significantly the specific antibody titer. The titer and avidity of the antibody raised may also be increased by admixture of the immunogen with a suitable adjuvant.

Haptens should be labeled to carrier proteins to provoke an immune response. This carrier protein should also be foreign for the host to be recognized by the T cells. For most immunogens, T cell–B cell interaction is essential for antibody production (so-called T cell-dependent immunogens). The immunogen should contain epitopes, which may be haptens, and carrier determinants, both of which need to be recognized. For an extensive review of this topic, the reader is referred to Klein (1982) and Langone and Van Vanakis (1981, 1983).

Two basic, often ignored, rules of immunization are essential: (i) obtain preimmunization serum and (ii) never pool different antisera.

The animal species most often chosen are rabbits, goats, swine, sheep, or chickens, though monkeys, guinea pigs, or horses are used for particular purposes. It has been shown that about 70% of normal human blood donors have antibodies in their serum which react with immunoglobulins of guinea pig, goat, sheep, and cow. Therefore, rabbits are often the most suitable. Due to their convenient size, several may be injected to find good responders among them.

A most common procedure for immunization is to prepare the immunogen at 2–4 mg/ml in an isotonic salt solution and mix vigourously with an equal volume of incomplete Freund's adjuvant (water-in-oil emulsion, not vice versa, i.e., a drop placed in cold water should remain intact; Herbert, 1973). This preparation may be stored at -20°C for successive immunizations.

Alternatively, protein stained with Coomassie blue may be cut as bands from polyacrylamide gels and used directly as immunogen (Boulard and Lecroisey, 1982). The dye does not interfere, but care should be taken to destain by diffusion and not by electrophoresis to prevent contamination of the immunogen. It is also possible to use immune precipitate obtained in agarose gels by suitable methods (diffusion, immunoelectrophoresis). This two-step procedure makes it possible to eliminate contaminants which copurify with the immunogen but do not coprecipitate in agarose, though this procedure evidently takes longer than direct immunization.

The dose of the immunogen is very important. Very little is gained in titer if more than 25–50 μg/kg of a substance of average immunogenicity is injected, but minor contaminants may elicit significant titers of unwanted antibodies. The minimum amount of an average immunogenic protein is about 20 μg/kg.

There is a tendency to overcharge the immunization regimen. Little is gained by injecting an immunogen with incomplete Freund's adjuvant more frequently than once a month. However, prolonging immunization for up to 1 year is beneficial for avidity. Rabbits, at least 3 months old, are injected subcutaneously or intracutaneously in the thick part of the skin above the shoulder blade.

Bleeding of the animal has been described in detail by Herbert (1973). A convenient method using the central artery of the ear has been described more recently by Gordon (1982). For a good blood flow, bleeding should take place in a relatively warm room (21–24°C; Harboe and Ingild, 1983).

For the production of monoclonal antibodies, the reader is referred to *Methods in Enzymology,* Volume 92 (published by Academic Press, Inc.). This large field is in a state of rapid evolution.

B. ISOLATION OF IMMUNOGLOBULINS

Isolation of immunoglobulin (Ig) is, compared to antibody purification, relatively simple. It is highly recommended for conjugation in cases in which high background levels in the assays are to be expected due to the presence of lipoproteins, hormone-binding proteins, or other serum factors. It also makes it possible to increase the titer by a concentration of about 10-fold than is possible with unpurified antiserum.

1. Salt Precipitation of Immunoglobulins

The most popular salts used for selective Ig precipitation are ammonium and sodium sulfate. The concentration of sodium sulfate is expressed as percentage (w/v), whereas the concentration of ammonium sulfate is expressed as percentage of saturation. The concentration of salt at saturation depends on the temperature, particularly for sodium sulfate [5× less at 4°C than at room temperature; for ammonium sulfate this varies less (3.9–4.1 *M*)]. This shows that the common expression in percentage is awkward, and that molarities are preferred.

The isolation of mammalian IgG and IgA by ammonium sulfate precipitation is usually achieved by two different approaches depending on the volume of the serum, i.e., for large volumes solid salt is directly added, whereas for small volumes salt is added as a concentrated solution (preferred).

For large volumes (Harboe and Ingild, 1983), 25 g ammonium sulfate is added slowly, while stirring, to 100 ml antiserum, and left for at least 3 hr at room temperature. The precipitate, which contains at least 98% of the immunoglobulin, is collected by centrifugation (4000 *g*, 30 min). The supernatant is discarded and the precipitate is washed with 25 ml 1.75 *M* ammonium sulfate. The remaining precipitate is collected again by centrifugation. This washing procedure can be repeated several times.

For small volumes, a saturated ammonium sulfate solution is first prepared by adding about 800 g salt to 1 liter of distilled water at 40°C. After dissolving for several hours and cooling overnight to room temperature, the pH is adjusted to neutrality. One volume of this saturated salt solution (separated from remaining solid) is added slowly, under continuous stirring, to two volumes of antiserum and the solution is equilibrated for 1 hr. The precipitate is collected by centrifugation and washed with a volume, equal to the original total volume, of 1.5 *M* ammonium sulfate as described above.

Contaminating lipoproteins can be removed by dialysis of the precipitate (taken up in distilled water), twice for 12 hr against distilled water at 4°C, once for a full day against 50 m*M* sodium acetate plus 21 m*M* acetic acid (pH 5.0), and repeated again in the same sequence (2× for 12 hr with distilled water and 1× for 24 hr with acetate). The precipitate formed during dialysis is removed by centrifugation and the supernatant is dialyzed, if necessary, against a buffer.

IgY (avian counterpart of mammalian IgG) can be isolated rapidly from chicken sera with sodium sulfate precipitation at room temperature. An equal volume of a 34% (w/v) salt solution is added to the serum, and the precipitate is collected by low-speed centrifugation and washed twice with 17% sodium sulfate. The precipitate is then redissolved in 0.06 *M* Tris–HCl, pH 8.0, and the IgY precipitated again by adding sodium sulfate to 150 g/liter. This cycle is repeated once and the salt is removed as described above.

2. Purification of Immunoglobulins by Ion-Exchange Chromatography

After salt precipitation, IgG can be further purified conveniently on DEAE-cellulose (Levy and Sober, 1960) DEAE-Sephadex A-50 (Harboe and Ingild, 1983), or DEAE-Sephacel.

The theory and application of ion-exchange chromatography have been discussed by Peterson (1970). Since then, several improvements have been made with respect to the nature of the carrier of the ion-exchange groups (e.g., microgranular cellulose, Servacel, Sephacel).

In the method of Harboe and Ingild, 1 g DEAE Sephadex A-50 is suspended in 50 ml 25 m*M* sodium acetate and left overnight. The buffer is then discarded, and a column is prepared and washed with 25 m*M* sodium acetate plus 21 m*M* acetic acid buffer, pH 5.0 (see Chapter 2, Section A). The sample is run over this column and eluted with the same buffer (60 ml/h). IgA and IgG are not retained and can be recovered almost quantitatively.

Alternatively, the IgG fraction obtained by salt fractionation is dialyzed against 17.5 m*M* phosphate buffer, pH 6.5, and passed on a column (e.g., DEAE-Sephadex) equilibrated with the same buffer. IgG passes directly except for a relatively small fraction of (electrophoretically fast-moving) IgG which can be eluted at pH 8.0, albeit with some contaminants (Pharmacia Product Information).

Another method was described by Tijssen and Kurstak (1974) using QAE-Sephadex A-50. This exchanger is equilibrated with ethylenediamine buffer [2.88 g/liter ethylenediamine and acetic acid (73 ml of 1 mol/liter)] pH 7.0. IgG will pass through immediately while contaminating proteins are retained. The latter can be desorbed with a buffer containing 435 ml 0.6 *M* acetic acid and 130 ml 0.6 *M* sodium acetate per liter (pH 4.0).

In each of these methods overloading should be avoided, since this will result in a spillover of the contaminants into the final sample.

Rowe and Fahey (1965) described a method, using DE-52, which permits the separation of several immunoglobulin classes. Proteins are adsorbed with a buffer of 5 m*M* phosphate, pH 8.0. A linear gradient (10–300 m*M*) is then applied. Three peaks emerge, i.e., an IgG containing fraction at 20 m*M*, whereas IgM and IgA are desorbed at higher salt concentrations. Tracey *et al.* (1976) and Capra *et al.* (1975) described variations of this method.

3. Isolation of IgG on Protein A-Sepharose

Protein A has an affinity for the majority of mammalian IgGs (Langone, 1982) and can be used to isolate these IgGs. Goding (1976) reported a simple method in which antiserum is applied on a 5 ml protein A-Sepharose column equilibrated with phosphate-buffered saline (PBS). After all nonbound proteins have passed, IgG is desorbed by applying a 0.9% sodium chloride solution containing 0.58% acetic acid. Protein A conjugated through *N*-hydroxysuccinimide esters to Affi-gels (Bio-Rad) seem to be more stable and to have more reactivity than the protein A-Sepharose.

It is also possible (Ey *et al.*, 1978) to elute IgG subclasses selectively by applying a buffer gradient with decreasing pH (0.1 *M* phosphate and 0.1 *M* citrate buffers). Human IgG_3 is not bound (Duhamel *et al.*, 1979). About 30% of IgA may also be adsorbed (van Kamp, 1979). The eluates and the column should be neutralized immediately after use.

4. Immunosorbents

Results with immunosorbents for the purification of specific antibodies for EIA are generally disappointing. It is difficult to recover high-avidity antibodies, whereas for EIA the high-avidity antibodies are required. However, immunosorbents remain convenient for the depletion of antibodies to contaminants.

A large number of matrices can be used, such as those with vicinal glycols which can be subjected to periodate oxidation (Ferrua *et al.*, 1979), cyanogen bromide-activated Sepharose (Livingstone, 1974), *N*-hydroxysuccinimide derivatized agarose (Maze and Gray, 1980).

Commercial CNBr-activated Sepharose is washed on a glass filter with 2 m*M* HCl (200 ml/g). Just before use, coupling buffer (0.1 *M* sodium bicarbonate, pH 8.2, containing 0.5 *M* sodium chloride) is added. The immunoreactant to be coupled is also equilibrated with this coupling buffer. The protein solution is gently mixed with the gel [protein:gel, 1:30 (w/w)]. If small haptens are used, the ratio should increase from 30 to 600. Mixing is continued for at least 3 hr (or overnight). The supernatant is removed and the remaining active groups are blocked by adding 0.2 *M* glycine or 1 *M* ethanolamine. The gel is washed alternately with pH 4.0 and pH 8.0 buffers. If small haptens are to be coupled, it is advantageous to use AH- or CH-Sepharose 4B to prevent steric inhibition of the antibody–hapten interaction.

N-Hydoxysuccinimide-derivatized agarose (available from Bio-Rad) is a very convenient alternative. The gel is washed on a glass filter with 3 volumes of isopropanol and 3 volumes of cold, distilled water (within 20 min). The ligand in a 0.1 *M* sodium bicarbonate solution is added and mixed with the gel for 1 hr at room temperature. Blocking is achieved by incubation with ethanolamine (0.1 ml 1 *M*) for another hour. The gel is then washed and ready for use.

Antisera are applied in neutral buffers at room temperature. The inclusion of 0.5% Tween 80 decreases possible nonspecific interaction with the gel matrix. To desorb adsorbed antibodies, chaotropic ions, organic acids with low surface tension, pH extremes (e.g., 2.5), or combinations thereof should be used depending on the nature of the antigen–antibody interaction. O'Sullivan *et al.* (1979) preferred 0.1 *M* NaCl-diluted HCl, pH 2.0, for proteins and 7 *M* guanidine-diluted HCl, pH 2.0, for small hydrophobic haptens.

5. Purification of IgY from Egg Yolk

Egg yolk contains, in contrast to egg white, large amounts of antibodies (IgY). An immunized chicken will produce about 100 mg antibody per egg. The purification, according to Jensenius *et al.* (1981), is quite simple. Yolk, separated from egg white, is diluted with 9 volumes of water and the pH is adjusted by adding 0.1 *N* NaOH. IgY is quite sensitive to reductive dissociation at slightly alkaline or neutral pH (Benedict and Yamaga, 1976). After freezing and thawing the lipids are removed by centrifugation, and IgY is further purified by sodium sulfate precipitation as detailed in Chapter 2, Section B,1.

C. PREPARATION OF Fab FRAGMENTS

Fab fragments have considerable advantages over the complete IgG, in EIA, both because of background levels and assay design (Chapter 4, Section A). Moreover, conjugates prepared with Fab penetrate more easily into tissue, and are, therefore, often applied in enzyme immunohistochemistry.

The IgG and their subclasses from the various species have different sensitivities to proteolytic cleavage.

Papain is used in 0.2 *M* sodium phosphate buffer, pH 7.4, containing 2 m*M* EDTA and 10 m*M* cysteine at 10 mg/ml. Papain is added to human or rabbit IgG (1 mg/100 mg IgG) and incubated at 37°C for 4–48 hr. The reaction is stopped by adding 10 m*M* iodoacetic acid. Alternatively, pepsin is dissolved in 0.2 *M* sodium acetate buffer, pH 4.7, and added to the IgG in the same buffer (1 mg/100 mg IgG) and left for 1–48 hr at 37°C. Sometimes a higher enzyme concentration is required (e.g., mouse IgG_{2a}), or a lower pH (mouse IgG_1; Lamoyi and Nisonoff, 1983).

Trypsin is the best suited protease for sheep IgG (Davies *et al.*, 1978). It is used at 2 mg/ml in 0.1 *M* Tris–HCl buffer, pH 7.8, containing 20 m*M* $CaCl_2$ and added 1 mg/100 mg IgG. One hour at 37°C suffices.

Fab fragments are purified by gel filtration, protein A, or on DEAE ion exchangers. The latter, in 5 m*M* sodium phosphate buffer, pH 8.0, will either let the Fab pass or retain it very slightly (desorbed by slightly increasing the NaCl concentration). The purity of the fragments is best verified by sodium dodecyl sulfate-polyacrylamide gel electrophoresis (SDS–PAGE).

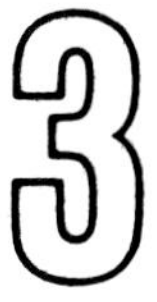

Conjugation of Enzymes

Conjugation procedures have been vastly improved in the last 15 years. Early procedures were based on chemical linkage of the enzyme to macromolecules by bifunctional reagents such as glutaraldehyde. The efficiency of these methods was poor. At most a small percentage of peroxidase was coupled. Better conjugation yields could be achieved for alkaline phosphatase, but this was accompanied by extensive polymerization. Current efficient methods are usually applicable or best suited for one or a few enzymes only.

Three different approaches exist. Chemical labeling is still the most popular, and some are very efficient. Immunological labeling takes advantage of the antigenicity of enzymes. Antibody reacting with this enzyme can be linked to the immune complex by an anti-immunoglobulin antibody and is useful. The enzymatic activity is not abrogated. Third, auxiliary nonimmune recognition systems (biotin/avidin, protein A, lectins) may be used to link the enzyme to the immune complex (bridge methods). In general, the latter two approaches result in higher detectability but more time-consuming designs. Sometimes a hybrid of these techniques can be used, such as the antibody–chimera technique (Porstmann *et al.*, 1984), in which primary antibodies are linked chemically to anti-enzyme monoclonal antibodies. The initial large difference between immunoassay detectabilities using chemically or immunologically enzymes vanishes almost completely.

To obtain optimal results, several important factors should be taken into account for chemical labeling. The law of mass action dictates the length of the incubation period required to obtain a certain amount of conjugate. Decreasing the concentration of the proteins to half its original value will require the prolongation of the incubation time by four times to obtain the same amount of conjugate (Kurstak *et al.*, 1984; Tijssen and Kurstak, 1984). At relatively low molar concentration ratios of enzyme/antibody, the distribution of enzyme molecules over the antibodies can be expected to follow a Poisson distribution. It can be calculated that at most 36.8% of the antibodies will be conjugated with one enzyme only (Tijssen, 1985). Increasing the molar concentration ratio beyond this optimum will result in an increase of the fraction of antibodies with two enzyme molecules, and a decrease in the fraction of free antibodies or 1:1 conjugates. Large conjugates are to be avoided for enzyme immunohistochemistry (problems with penetration of fixed tissue and background staining). Moreover, the biologic activity of enzymes or antibodies in large complexes is usually severely impaired.

The optimum ratio of the molar concentrations of activated enzyme and antibody is around 1. It should be realized that sometimes only a small fraction of the enzyme is activated (e.g., less than 10% of peroxidase with a large excess of glutaraldehyde). Therefore such procedures are extremely wasteful for the expensive enzyme. The excess of activated peroxidase to be added to obtain a certain molar ratio of bound enzyme/IgG is, however, small (Fig. 3.1).

The relative reactivity of some cross-linkers to proteins may also be very different. Modesto and Pesce (1971) noted that 4,4′-difluoro-3,3′-dinitrophenyl sulfate (FNPS) reacts 66 times faster with IgG than with peroxidase. In an equimolar mixture predominantly IgG polymers are formed, to some of which peroxidase may be attached. The reactivity of these cross-linkers is also determined by the solubility and the stability of these agents, and, often, the pH.

In this overview, the most satisfactory procedures for conjugation are selected. Moreover, we present the purification of some of the key enzymes in enzyme immunoassays. The enzymes can be prepared by simple methods at a small fraction of the commercial cost, and often have an activity considerably higher than those obtained commercially. Conjugation in the laboratory is recommended since it is considerably cheaper, better reagents are usually obtained, and any antibody, antigen, or enzyme may be conjugated (not dependent on commercial availability).

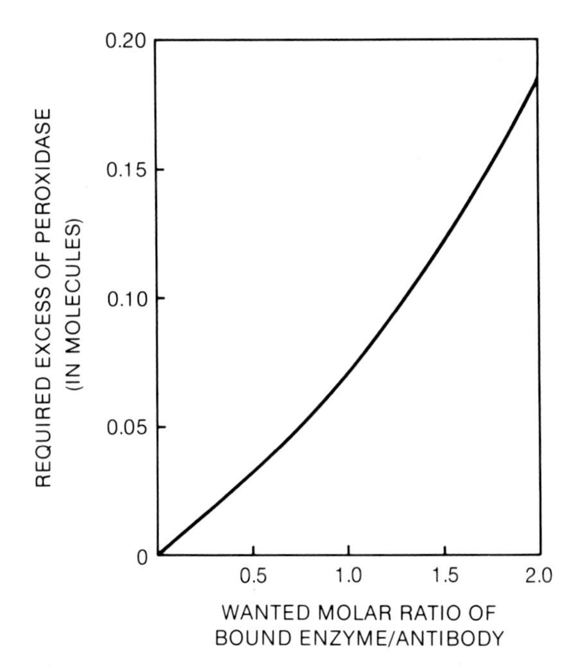

Fig. 3.1. The excess of activated peroxidase required to obtain a certain average molar ratio of bound enzyme per antibody. For example, if a molar ratio of 2.0 for peroxidase/IgG is desired, then an excess of 0.18 molecules, or a total of 2.18 molecules of activated peroxidase should be added to each molecule of IgG.

A. PURIFICATION OF ENZYMES

1. Horseradish Peroxidase

Commercial "pure" peroxidase contains, beside the highly active "C" isozyme, less active isozymes and colorless inactive proteins. Purification of peroxidase by a simple, single-step method (Kurstak *et al.*, 1984; Tijssen and Kurstak, 1984) from crude peroxidase extracts (e.g., Sigma, type II) yields peroxidase of high purity and activity, and is 5–10 times cheaper. Peroxidase thus prepared is better suited for enzyme immunoassays than the "pure" commercial peroxidase (e.g., Sigma, type VI).

Purity of peroxidase is expressed generally in RZ (Reinheits Zahl) numbers. RZ equals the absorbance at 403 nm/275 nm (at 403 nm the Soret band of the hemin group, and at 275 nm the protein). Though it is assumed that an RZ of 3.0 indicates pure enzyme, this is usually not true. In fact, the 7 isozymes of horseradish peroxidase all have different RZ values when pure (from below 3 to above 4; Kurstak *et al.*, 1977; Kurstak, 1985).

The RZ of the major and most active component of commercial "pure" peroxidase is 3.5 (Kurstak *et al.*, 1984; Tijssen and Kurstak, 1984) instead of the 3.0 of that mixture. The purification method consists of a single step by which all contaminants are retained on an ion-exchange column, whereas the highly active "C" isozyme passes freely. Virtually all activity can be recovered.

Crude peroxidase is dissolved in 1.5 m*M* sodium phosphate buffer, pH 8.0, and a DEAE-Sepharose column is equilibrated with the same buffer. Up to 5 mg is applied per ml gel. Pure peroxidase passes directly and is recovered (monitoring at 403 nm or visually). The RZ of this preparation is about 3.25–3.45. If "pure" peroxidase (Sigma type VI) is passed on such a column an RZ of around 3.50 is obtained. The specific activity of this enzyme is about 20–25% higher than for the commercial pure enzyme.

2. Alkaline Phosphatase

This extremely expensive enzyme can be purified by a simple affinity-chromatography method (Mössner *et al.*, 1980).

A column is prepared after equilibration of L-histidyldiazophosphonic acid-agarose (from Sigma) with Tris–HCl buffer, pH 8.0. The enzyme, dialyzed against several changes of this buffer over a 24-hr period, is applied (100 U/ml gel). The adsorbed enzyme is eluted by adding 20 m*M* disodium phosphate. It is important to dialyze the enzyme, subsequently, against the Tris buffer containing 0.1 m*M* zinc chloride and 1 m*M* magnesium chloride; the column is reequilibrated with the Tris–HCl buffer.

3. Microperoxidase

Plattner *et al.* (1977) reported a simple method of purification for the extremely expensive microperoxidase (the proteolytic fragment of 9 amino acids containing the active site of cytochrome *c*). Cytochrome *c* is digested by passing the enzyme slowly through a long, narrow column (50 × 0.52 cm) of trypsin-Sepharose (e.g., from Sigma) in 0.1 *M* ammonium carbonate buffer, pH 8.5, at room temperature (flow rate 0.5 cm/hr). The red-colored fraction is lyophilized and purified on Sephadex G-50 (150 × 0.8 cm; 5 ml/hr). Sephadex interacts with the heme nonapeptide, resulting in a delay of the passage of this fragment (elution after about 40 ml). The fragment thus obtained is 99% pure.

Similarly heme octapeptide can be prepared by a pretreatment of cytochrome with pepsin in 0.1 *N* HCl for 24 hr (Tijssen and Kurstak, 1974), prior to the passage over the trypsin-Sepharose and Sephadex columns as described above.

B. SELECTED PROCEDURES OF CONJUGATION

1. Periodate Method for the Coupling of Peroxidase and Glucose Oxidase

Nakane and Kawaoi (1974) described an elegant method for the efficient labeling of glycoproteins to other macromolecules. The method is based on the generation by periodate oxidation of aldehydes of vicinal glycol groups of the carbohydrates. These aldehydes form Schiff bases with the amino groups of other proteins. The Schiff bases can be stabilized by treatment with a reducing agent (sodium borohydride).

The original method, though very efficient, resulted in considerable inactivation of the enzyme and the formation of undesired polymers (Nygren, 1982). Several improvements and simplifications of this technique have been reported (Wilson and Nakane, 1978; Kurstak *et al.*, 1984; Tijssen and Kurstak, 1984).

Central in this procedure is the oxidation of the carbohydrate moiety. Increasing the degree of oxidation will result in a greater number of aldehyde groups up to a certain point. Too little oxidation prevents an efficent conjugation, whereas a strong oxidation results in the formation of carboxyl groups instead of aldehyde groups. Thereafter, relatively more carboxyl groups will be formed, since the aldehydes are more sensitive to oxidation than are the vicinal glycol groups. Moreover, at high periodate concentrations amino acids may be oxidized, rendering them more hydrophilic and thus affecting the conformation of the enzyme (de la Llosa *et al.*, 1980). Current methods often tend to overoxidize the enzyme (5–10 times too much periodate). Care should be taken to perform the oxidation in a defined buffer and not in distilled water since this oxidation is highly pH dependent (Dyer, 1956). Incubation should be in the dark.

The simple method from our laboratory (Kurstak *et al.*, 1984; Tijssen and Kurstak, 1984) includes four steps: activation of peroxidase, conjugation, stabilization, and purification of conjugates. For activation, 5 mg of peroxidase (11 OD_{403} units) in 0.5 ml freshly prepared 0.1 *M* sodium carbonate is mixed with 0.5 ml of 8–16 m*M* sodium *m*-periodate and incubated at room temperature for 2 hr in the dark in a closed tube. For conjugation, 15 mg IgG (in 1–2 ml 0.1 *M* sodium carbonate buffer, pH 9.2) is added to the activated peroxidase. This mixture is added to a sealed (by flaming) Pasteur pipet, plugged with glass wool, and dry Sephadex G-25 is added (weight corresponding to one-sixth of the combined weight of peroxidase and IgG). The instant swelling of Sephadex will cause an increase in the protein concentrations and at the same time the periodate still present will be consumed by the Sephadex. After a 3-hr incubation, the conjugate is eluted from the Sephadex and mixed with 1/20 and 3/20 columns of freshly prepared sodium borohydride (5 mg/ml in 0.1 m*M* sodium hydroxide) at 30 min intervals. For the purification of the conjugate, IgG and IgG-peroxidase is first precipitated with 50% ammonium sulfate (leaving free peroxidase in the supernatant). The pellet obtained by centrifugation is dissolved and equilibrated with PBS. Con A Sepharose, also equilibrated with PBS, retains the IgG-peroxidase from the conjugation sample, whereas free IgG passes directly. The conjugate is then desorbed by adding PBS containing 0.05 *M* α-methyl-D-mannopyranoside. It is advantageous in these procedures to raise the sodium phosphate concentration to 0.1 *M* to decrease nonspecific interaction of the proteins with the Sepharose.

The RZ of the conjugate of satisfactory conjugates generally is between 0.3 and 0.4. This method can be successfully applied for glucose oxidase as well.

2. Glutaraldehyde for the Conjugation of Alkaline Phosphatase and for the Two-Step Procedure of Peroxidase

For the conjugation of alkaline phosphatase, the enzyme and antibody solutions are extensively dialyzed against 0.1 *M* phosphate buffer, pH 6.8. Solutions of 1 ml of enzyme (10 mg) and 1 ml of IgG (5 mg) or Fab (2.5 mg) are mixed and 0.05 ml of a 1% glutaraldehyde solution is added (Avrameas *et al.*, 1978). After an incubation of 3 hr at room temperature, 1/20 volume of 1 *M* L-lysine is added to block reactive sites. This preparation is then extensively dialyzed against Tris-buffered saline.

A frequently used method to obtain 1:1 conjugates of peroxidase and IgG (or Fab) is based on the observation that peroxidase reacts poorly with glutaraldehyde. Therefore, an excess of glutaraldehyde (0.2%) is added to peroxidase (10 mg in 0.2 ml of 0.1 *M* phosphatase buffer, pH 6.8) and left for 18 hr. The excess of glutaraldehyde is then removed by dialysis against 0.9% sodium chloride (or by passage over Sephadex G-25). Subsequently 5 mg IgG or 2.5 mg Fab and 0.2 ml of 0.5 *M* sodium carbonate, pH 9.5, are added and left for 24 hr at 4°C. The remaining active groups are blocked by adding 0.1 ml of 1 *M* lysine for 2 hr and the sample is neutralized to pH 7.0. This preparation is then extensively dialyzed against PBS.

3. The *N,N'-o*-Phenylenedimaleimide-Mediated Coupling of β-Galactosidase

N,N'-o-Phenylenedimaleimide is an excellent cross-linker for macromolecules possessing sulfhydryl groups (Hamaguchi *et al.*, 1979). IgG or $F(ab')_2$ may be reduced to generate sulfhydryl groups by dialysis of the antibodies against 0.1 *M* sodium acetate buffer, pH 5.0, and, after dagassing, adding slowly 0.1 ml of 0.1 *M* 2-mercaptoethylamine. It is then incubated for 90 min at 37°C under nitrogen. The reducing agent should be removed prior to conjugation, while keeping the same buffer. A saturated solution (0.75 m*M*, equal volume) of cross-linker is added and incubated for 20 min at 30°C. Excess of cross-linker is removed by gel filtration through Sephadex G-25 (in 0.02 *M* sodium acetate, pH 5.0). After concentration (under reduced pressure) to 0.3 ml, the pH is adjusted to 6.5 with 0.25 *M* sodium phosphate buffer, pH 7.5. Twenty microliters of *N*-ethylmaleimide-BSA, 1 μl of magnesium chloride, and 0.1 ml of β-galactosidase (5 mg/ml) are successively added and the preparation is incubated for 16 hr at 4°C. The volume is then adjusted to 1 ml by adding PBS containing 1 m*M* magnesium chloride, 1 mg/ml BSA, and 0.1% sodium azide (to destroy residual maleimide).

4. Bissuccinic Acid *N*-Hydroxysuccinimic Ester for the Conjugation of 8-Microperoxidase

The method of Ryan *et al.* (1976) is useful for 8-microperoxidase since the latter has only one reactive group, which can be completely converted by this cross-linking. The excess of this bifunctional reagent is removed prior to the addition to Fab.

In detail, 5 mg of cross-linker is dissolved in 300 μl dimethyl sulfoxide.

The microperoxidase is dissolved separately in 250 μl pyridine and this solution is added very slowly (dropwise over 2.5 hr) to the cross-linker solution. After an additional 2.5 hr incubation at room temperature, 4 volumes of diethylether are added to allow the product to precipitate overnight (at 4°C). The supernatant is carefully removed and the precipitate is dissolved in 400 μl dimethyl sulfoxide. This activated 8-microperoxidase is stable for at least a year at 25°C.

Activated microperoxidase is added slowly (over 90 min) to IgG or Fab (15 mg/ml) to a maximum of one-eighth of the volume of the latter and left for 4 hr at room temperature and overnight at 4°C, respectively.

The purification consists of two steps: gel filtration (BioGel P300 2.5 × 70 cm; 0.1 *M* Tris–HCl buffer, pH 7.4) to remove the small reagent or by-products; and ion-exchange chromatography to eliminate unsubstituted antibody. For the second step, DEAE-cellulose is equilibrated with 0.01 *M* sodium phosphate buffer, pH 7.4. Free antibody passes directly, while the substituted fraction is eluted by adding 0.3 *M* potassium chloride to the buffer. The substitution rate will be about 1.5, but may be increased by changing the relative amounts of activated microperoxidase and antibody solutions.

5. Coupling of β-Galactosidase with *m*-Maleimidobenzoyl-*N*-Hydroxysuccinimide Ester

This reagent has an advantage over *N,N'-o*-phenylenedimaleimide; only one of the two macromolecules needs to possess free sulfhydryl groups. Thus β-galactosidase can be directly conjugated to IgG without a reduction step. The NHS ester will acylate the free amino group, whereas the maleimide residue will form thio ethers with sulfhydryl groups.

For the conjugation of β-galactosidase, to each milliliter containing 1 mg/ml of IgG in 0.1 *M* phosphate buffer, pH 7.0, and 0.05 *M* sodium chloride, 20 mg MBS/ml in 10 μl of dioxan is added and incubated for 1 hr at 25°C. In the original procedure 30°C was used, but degradation of MBS can then be extensive. The excess of reagent is subsequently removed by filtration on Sephadex G-25 equilibrated with 0.1 *M* phosphate buffer, pH 7.0, containing 20 m*M* magnesium chloride and 0.05 m*M* sodium chloride. The antibody containing fractions are pooled, mixed immediately with β-galactosidase (equal amount, in milligrams, to IgG) and incubated for 1 hr at 30°C. The reaction is stopped by adding a reducing agent (e.g., 0.01 *M* 2-mercaptoethanol).

6. 4-(*N*-Maleimidomethyl) Cyclohexane-1-carboxylic Acid *N*-Hydroxysuccinimide Ester (MCC–NHS)

MCC–NHS is a superior cross-linker for peroxidase, glucose oxidase, and alkaline phosphatase (Ishikawa *et al.*, 1983), but not for β-galactosidase. It is much more stable than the other maleimide derivatives. The principles of its use resemble those of MBS, i.e., one of the macromolecules should contain a thiol group.

MCC–NHS should be dissolved in *N,N'*-dimethylformamide (1.6 mg/20 μl) at 30°C. Enzyme (2 mg peroxidase or 0.4 mg glucose oxidase) is dissolved in 0.3 ml of 0.1 *M* sodium phosphate buffer, pH 7.0, and 20 μl of the cross-linking agent is added. The mixture is gently stirred for 1 hr at 30°C. Sometimes a precipitate may form, which should be removed by centrifugation. The enzyme is subsequently passed through a Sephadex G-25 column, equilibrated with 0.1 *M* phosphate buffer, pH 6.0. The enzyme fractions are concentrated in the cold, and thiolated antibody is added (3 times molar excess). After an incubation for 24 hr at 4°C, mercaptoethylamine is added to 1 m*M* to destroy the remaining maleimide groups. The conjugates are purified by gel filtration or affinity chromatography (Chapter 2, Section B,4).

7. Immunological Conjugations

The activity of some enzymes, such as peroxidase, is not affected after reaction with anti-enzyme antibodies from polyclonal sera. In the other cases, monoclonal antibodies can be selected which do not affect activity.

In the original, "unlabeled antibody," method the reagents are applied sequentially, i.e., after the primary antibody has reacted with antigen, an excess of anti-immunoglobulin antibody is applied. This excess ensures that one of the antigen-binding sites remains free, which can capture anti-enzyme antibodies, produced in the same animal as the primary antibody. These, in turn, will subsequently immobilize added enzyme. The weak point in this time-consuming sequence is the anti-enzyme antibody. Since this fraction seldom contains only anti-enzyme antibody, many nonspecific antibodies are captured by the anti-immunoglobulin antibody leading to poor detectability. The purification of anti-enzyme antibodies seldom improves results in the case of peroxidase, since high-avidity antibodies are lost in this procedure (Sternberger, 1969). For alkaline phosphatase, however, this approach is recommended (Mason and Sammons, 1978; Mason *et al.*, 1983). This problem can be overcome by preforming soluble enzyme–anti-enzyme complexes, or by the use of monoclonal antibodies.

Preformed peroxidase–anti-peroxidase (PAP) complexes (Sternberger *et al.*, 1970) are very convenient. The anti-peroxidase antibodies from an antiserum are precipitated by adding the enzyme. The precipitate is collected by low-speed centrifugation and washed with cold saline. A solution of peroxidase (4 mg/ml) is added until a 4-fold excess of peroxidase is obtained. The pH is brought to 2.3 by adding HCl. The dissolved precipitate is neutralized with sodium hydroxide. The solution is cleared by centrifugation and an equal volume of saturated ammonium sulfate is added (equilibration time, 25 min). This will precipitate only the PAP complexes, whereas free peroxidase remains in solution.

The PAP complexes are rather large (about 420,000 Da) and consist usually of 3 peroxidase and 2 IgG molecules. This limits their use for enzyme immunohistochemistry since penetration of the tissue may pose problems. For ultrastructural studies, this problem can be overcome since immunological localization can be performed on thin sections. However, the large complexes and the use of several layers of the large antibodies result in a less accurate localization of the epitope.

Monoclonal antibodies are very convenient for preparing soluble enzyme–antibody complexes. They are smaller than the polyclonal complexes, since only one epitope on the enzyme can be recognized, and will consist of 2 enzyme molecules and 1 antibody. About 0.01 mg enzyme is added to undiluted culture supernatants. This material can be used after a 1000-fold dilution, without further purification.

Principles of the Design of Enzyme Immunoassays

A. FORMULATION OF METHODOLOGY

Enzyme immunoassays (EIA) take advantage of the biological properties of two important macromolecules, enzymes and antibodies.

Antibodies, which can be raised to almost any compound foreign to the host, are able to distinguish closely related compounds. They recognize the specific compound (hapten or antigen) by multiple interactions through complementary surfaces (affinity). The durations of the individual bindings are usually rather short (about 1 sec; Tijssen, 1985). However, multiple interactions reinforce the complex by several orders of magnitude, a phenomenon known as avidity. Enzymes are biological catalysts, which accelerate specific chemical reactions enormously and can thus be detected by adding substrate.

The strategy in EIA is to conjugate an appropriate enzyme to one of the immunoreactants, let the immunological reaction take place, add substrate, and determine the substrate–product conversion by the enzyme. The amount of product indicates how much enzyme-labeled immunoreactant is incorporated in the complex.

This approach has important advantages over radioimmunoassays (RIA), attaining similar detectability in competitive assays and higher detectabilities than RIA in noncompetitive assays. Whereas, in RIA each labeled species gives only one (decay) signal, in EIA an enzyme continues to build up a product, often at a rate of about 100,000 molecules per minute. Moreover, EIA does not require expensive equipment, EIA makes the use of dangerous radioactive materials unnecessary (though in some assays toxic materials are used), EIA does not require licensing, reagents for EIA are very stable unlike the expensive reagents of RIA, and EIA does not have the same degree of disposal problems as RIA. In addition, the antigenic character of enzymes increases the sophistication of the assays.

In general, two types of EIA can be distinguished, homogeneous (in solution) and heterogeneous (i.e., one of the immunoreactants is anchored to a solid phase). Homogeneous assays are most often used for the detection of small compounds (haptens), whereas the heterogeneous EIA usually are applied for the detection of large molecules or particles (e.g., infectious agents). Though these general types are frequently used for the classification of enzyme immunoassays, the assays should be classified according to the underlying principles. Analysis of the various EIA reported demonstrates the two groups of assays with essential differences can be discerned (Kurstak *et al.,* 1984; Tijssen, 1985; van Weemen, 1985). These were named Activity Amplification (AA) and Activity Modulation (AM) assays. In AA-type methods, the detector immunoreactant is used in large excess to detect the other. Due to the law of mass action, exceedingly small amounts of sample molecules can then be detected. However, the relative potencies of the antigens for the various antibodies will appear similar (due to the large excess) and, consequently, specificity of such methods may be poor. The contrary applies for AM methods in which low concentrations of immunoreactants are used so that sample molecules will affect more strongly the standard interactions. The detectability of such systems equals the relative experimental error divided by the affinity constant (Tijssen, 1985). For example, for a system with an affinity constant of 10^8 M^{-1} and a relative error of 10% the minimal amount that can be detected is $0.1/10^8$ or 10^{-9} M. The AA-type EIA makes it possible to detect sample molecules at significant lower concentrations, but in AM assays the specificity reflects directly the differences in the various affinity constants of the antibodies. Specificity is thus much better than for AA assays. AM assays are not strictly synonymous to competitive assays, since in AM assays AA steps may be included.

It is convenient, for this discussion, to differentiate the EIA designs into four groups (homogeneous/heterogeneous and competitive/noncompetitive).

1. Noncompetitive Solid-Phase Enzyme Immunoassays

Either antigens or antibodies may be detected. In cases of infectious diseases, antibodies are detected only after an incubation time, thereby allowing the perpetuation and spreading of the disease. Sensitive enzyme immunoassays may often permit the direct detection of antigens.

The immunoreactant, complementary to the molecule to be detected (e.g., antibody for the detection of antigen), is immobilized on the solid phase. This sensitized solid phase serves to extract the target molecules from the sample fluid. This stage is followed by one or more steps with antibodies, one or more of which are labeled with an enzyme. This is a typical AA assay (Fig. 4.1); increasing the detectability will decrease the specificity of the assay. Numerous variations can be designed for these AA assays, using antibody fragments or other proteins such as protein A (Fig. 4.2).

A recent variant of this assay, for which the same principles apply, is the class-capture assay (Fig. 4.3). The trapping agent (i.e., anti-IgG, anti-IgM, anti-IgA, etc.) is immobilized on the solid phase, followed by an incubation with the serum. The subsequently added antigen will be bound only if there are antibodies among the immunoglobulins of the particular class extracted from the serum. The quantity of antigen bound is directly related to the antibody titer in that immunoglobulin class and can be quantitated by enzyme tracers.

The underlying principles of the enzyme immunohistochemistry assays are similar to the titration assays of this group.

2. Homogeneous Noncompetitive Enzyme Immunoassays

Monoclonal antibodies with specificities against different epitopes on the antigen will on reaction with that antigen be in a proximal vicinity. When these two antibodies are labeled with two different enzymes so that the product of one will be the substrate for the other (e.g., hydrogen peroxide produced by glucose oxidase is consumed by peroxidase), the local concentration of the substrate is high for the second enzyme and the antigen is detected (Fig. 4.4). If the antibodies are not brought together by the antigen, they will be widely apart and the activity of the second enzyme

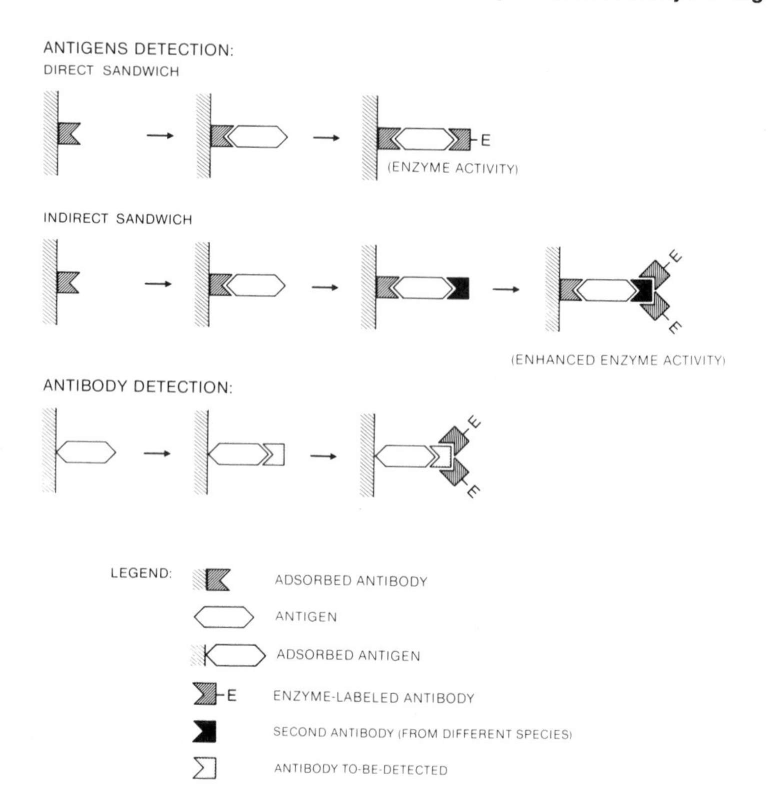

Fig. 4.1. Solid-phase enzyme immunoassays using the sandwich method. The solid phase is coated with specific antibody for antigen detection or antigen for antibody detection. The immunosorbents thus obtained capture the corresponding molecules from the sample added. Immobilized sample molecules are then detected with enzyme-conjugated antibody. As shown for the antigen detection, an indirect sandwich can be constructed for enhanced sensitivity.

will be very limited. This AA approach is still in the development stage (Ngo and Lenhoff, 1981; Kurstak, 1985; van Weemen, 1985).

3. Competitive Solid-Phase Enzyme Immunoassays

A large number of variants of these assays have been described. Competition occurs always in the second layer (AM step). In this step enzyme-labeled immunoreactants can be used, or omitted (as in mixed AA–AM assays). The molecule to be detected competes either with the labeled

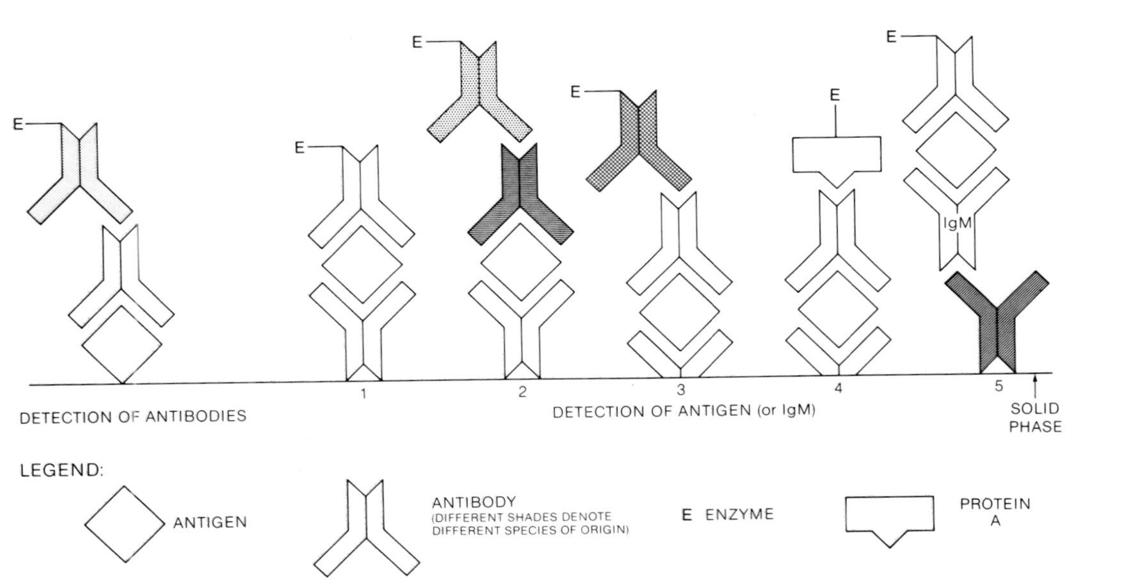

Fig. 4.2. Variations of solid-phase EIA used in immunodiagnosis. The labeled antibody in this diagram may consist of several layers of immunoreactants (bridge, double bridge, etc.). For the detection of antibodies, antigen is immobilized to serve as a trapping reactant whereas for the detection of antigens the antibody is immobilized. Most common is the direct sandwich method (1); if the indirect sandwich method is used antibodies should be obtained from two different species (2), or $F(ab')_2$ fragments immobilized so that the conjugate can only recognize the Fc fragment of the second antibody layer (3). The use of enzyme-conjugated protein A as a universal detection agent is particularly useful (4). In the class-capture technique (5), the macromolecule to be detected is also in the second layer (5). According to the design a higher detectability or higher specificity is obtained.

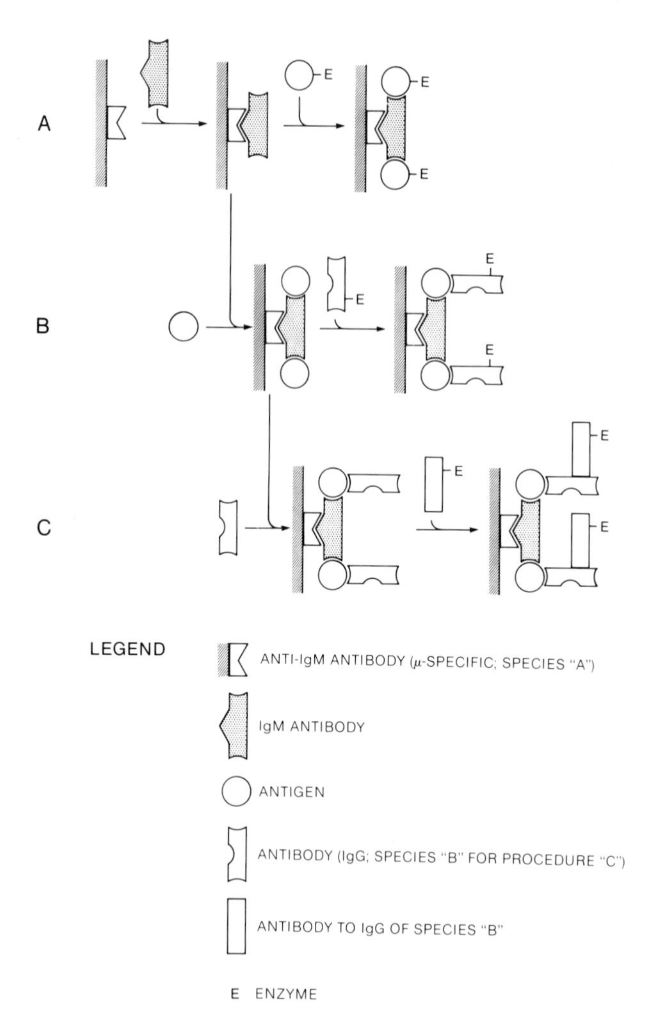

Fig. 4.3. Class-capture assay for the detection of IgM antibodies. An anti-IgM antibody (M specific) is immobilized on the solid phase. This immunosorbent will capture the IgM molecules present in the fluids to be tested. The presence of IgM can then be revealed by adding enzyme-labeled antigen (A), or directly by antigen and enzyme-conjugated antibody to the antigen (B) or by the indirect method (C), which is most sensitive but also most prone to nonspecific staining.

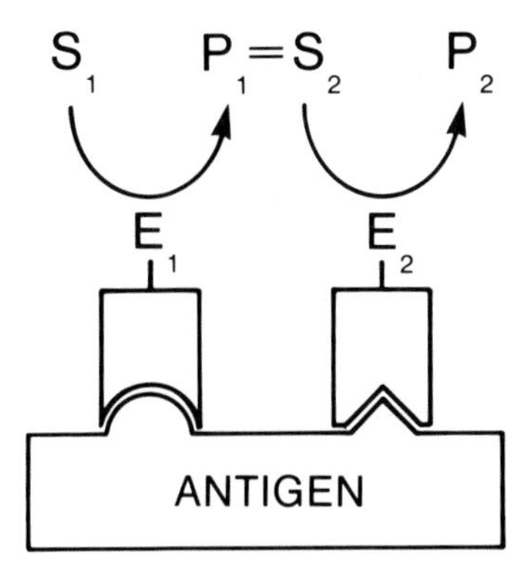

Fig. 4.4. Homogeneous activity-amplification enzyme immunoassays. The antigen in the solution has different epitopes, which bind different antibodies, each of which has been labeled with a different, preselected enzyme. The enzyme E yields a product which in turn is the substrate for enzyme E_2. If the two enzymes were distributed at random throughout the solution, the amount of substrate available for E_2 would be limited. However, the closeness of E_1, and E_2, brought about by the antigen if present, yields a significant amount of P_2 since the local concentration of S_2 would be much higher than the average concentration.

species for the solid-phase immunoreactant, or with the solid-phase immunoreactant for the labeled immunoreactant (Fig. 4.5).

It should be realized that either antigen or antibody can be labeled.

4. Homogeneous Competitive Assays

These methods can be based on two different phenomena: the reaction of the antibody with the hapten attached to enzyme, substrate, or cofactor may cause steric inhibition and thus modify the enzymatic activity or change the conformation of the enzyme resulting in a modulation of the enzyme activity.

The conformational change of the enzyme may increase or decrease the activity. It is not rare to have in the same antiserum a fraction of antibodies increasing activity and another fraction decreasing the activity (Rowley *et al.*, 1975).

Haptens present in the sample will bind with the antibody and thus prevent a proportional amount of antibody from reacting with the enzyme. This will decrease the enzyme modulation (Fig. 4.6).

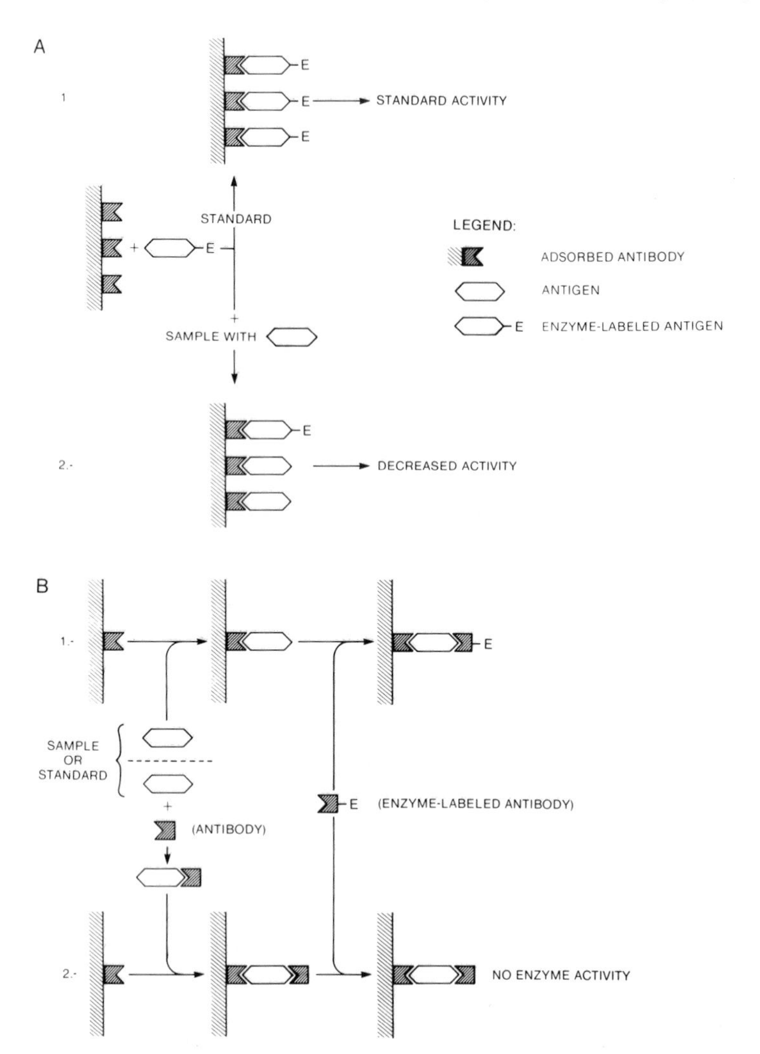

Fig. 4.5. Competitive solid-phase enzyme immunoassays. In scheme A, solid-phase coated antibody extracts enzyme-labeled antigen from a standard solution yielding a standard activity (A.1). If, however, the sample added to the conjugate also contains antigen, a competition between the two antigen species decreases the enzyme activity. In another approach, antibody can be added to the sample (B). If only the standard antigen is present, this will be neutralized and not be able to react with the conjugate. However, if antigen is present in the sample, this will compete with the standard antigen for the available antibody, resulting in less neutralization and higher enzyme activity on the solid phase.

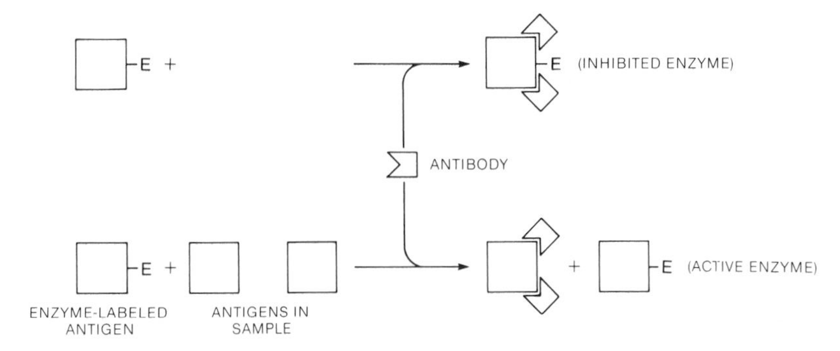

Fig. 4.6. Competitive, homogeneous enzyme immunoassays. This type of assay is used particularly for small molecules such as drugs, hormones, etc. If no antigen is present in the sample, the antibody will be able to react with the enzyme-conjugated antigen (hapten) and thus inhibit the enzyme. The presence of antigen in the sample will decrease the amount of antibody reacting with the conjugate, resulting in less inhibition.

B. SOLID PHASES USED IN ENZYME IMMUNOASSAYS

Solid phases, to which one of the immunoreactants is attached, enable a fast and complete separation of reacted immunoreactants from those in the supernatant by decantation or centrifugation. Many different solid phases can be used. Plastics are by far the most popular, in particular polystyrene and polyvinyl microtiter (12 × 8 wells) plates. They are convenient in that the separation steps are extremely simple and the amount of product in relatively small samples can be assessed by measuring the optical density through the plastic. However, plastics also have important limitations in that they are immunoreactant consumptive (e.g., require 10–1000 times more material than the nitrocellulose membrane solid phase) and often less than 10% of the immunoreactant added is actually adsorbed (Cantarero *et al.,* 1980). Moreover, the avidity seems to decrease 10–100 times on immobilization of the reactant, and the speed of antibody–antigen interactions decreases (from minutes to hours).

Nitrocellulose should prove very useful in qualitative assays or if very small samples (less than 1 μl) or ionic detergent-solubilized antigens are to be tested. Nitrocellulose binds close to 100% of the immunoreactants, but precise quantitation is possible only by reflectance densitometry.

Many particulate solid phases (e.g., beads used for immunosorbents: agarose, cellulose, polyacrylamide, dextran) are possible. The immunoreactants are bound covalently so that leakage, as in the above methods, is greatly reduced. Moreover, the solid phase is dispersed throughout the sample, accelerating considerably the rate of antibody–antigen interaction. The surface area on these beads is also much greater than those in wells on the other solid phases with comparable volumes. A disadvantage is the more cumbersome separation.

1. Plastic Solid Phases

Current methods, based on noncovalent, adsorption of one of the immunoreactants to the plastic, are simple, and generally give good results. However, desorption can be significant during the assay, e.g., Hermann *et al.* (1979) noted a loss of 68% of adsorbed antigen during the test. Desorption may be highly influenced by the serum used (Dobbins Place and Schroeder, 1980).

Protein adsorption to plastic is believed to occur through hydrophobic interactions (Cantarero *et al.*, 1980). The adsorption was shown to be independent of the net change of the protein, but each protein has a different binding constant. Since the speed of hydrophobic interactions increases with higher temperatures, it can be expected that higher temperatures are beneficial. This has been corroborated by studies in which noncovalent adsorption of a viral antigen was virtually complete after 10 min at 37°C (or 1 hr for IgG_0 and that more was bound than at the customary overnight incubation at 4°C (Fig. 4.7, Tijssen *et al.*, 1982; Kurstak *et al.*, 1984). Two phases can be distinguished in the adsorption process: protein–plastic interaction followed by protein–protein interaction (stacking) (Cantarero *et al.*, 1980). The latter phenomenon should be avoided, since this binding is not very stable and the stacked protein may easily desorb during the test and thus eliminate immunoreactants attached to them or compete for the immunoreactants in the sample or other incubation steps. There is, therefore, a definite optimum concentration for adsorption, which is different for each protein due to their inherent adsorptive properties. This corresponds usually to about 1–10 μg of immunoreactant per ml. The amount of protein bound to the plastic is usually about 1.5 ng/mm^2.

The plastics, on the other hand, may not be very uniform with respect to their adsorptive characteristics. Kenny and Dunsmoor (1983) observed that two groups of plates are encountered, those that adsorb albumin well and those that do not. The former are better suited for the adsorption of mixtures of immunoreactants. Variability among plates of the same batch can be significant. Most notable is the "edge effect" (Chessum and Denmark,

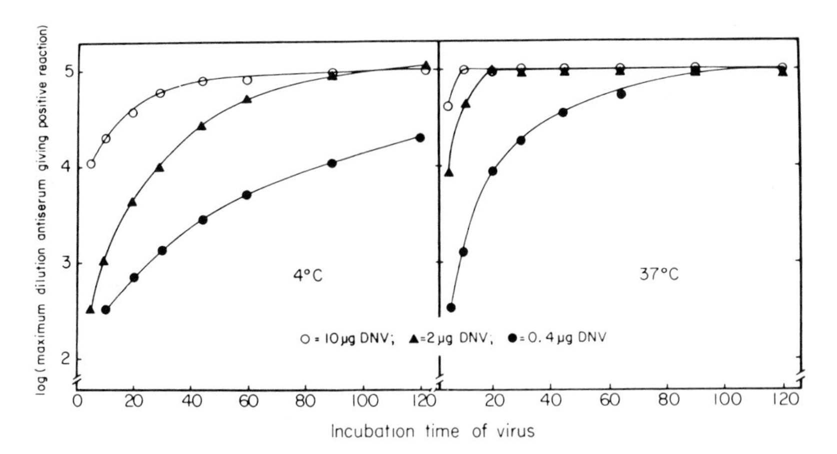

Fig. 4.7. Secondary plots of the results obtained with different conditions of sensitization of the polystyrene plates with densonucleosis virus (DNV). Three different concentrations of virus (0.4, 2, or 10 μg) were incubated for various periods of time at either 4 or 37°C. The dilution which gave a response of 0.15 OD/30 min was determined on a primary plot and plotted against the time on the secondary plot.

1978), which made many investigators decide not to use the outer wells. Several explanations of this effect have been given (Burt *et al.*, 1979), but it seems that at least a part of this problem can be traced to the poor conductivity of the plastic and to the fact that thermal gradients are generated from the outer wells to the center of the plate (Oliver *et al.*, 1981). Preincubation of the plates and the solutions used for coating at the desired temperature gave a much lower coefficient of variation.

The buffers used for the coating of the plastic wells with the immunoreactants are not critical for most cases. They should not contain detergents, and the pH should not be at the p*I* of the protein to be adsorbed to prevent protein stacking. Widely applied buffers are 50 m*M* carbonate buffer, pH 9.6, 20 m*M* Tris–HCl, pH 8.5, containing 0.1 *M* sodium chloride, or 10 m*M* sodium phosphate buffer, pH 7.2, containing 0.1 *M* sodium chloride. The most often used carbonate buffer may not be appropriate for all purposes. Barlough *et al.* (1983) observed that the use of this buffer for the adsorption of coronavirus antigens resulted in diffuse and nonspecific staining, whereas the use of the sodium phosphate buffer, 0.9% sodium chloride or just distilled water gave excellent results. It is, therefore, recommended that different coating buffers be tested for each new procedure.

If the immunoreactant to be coated is present in a mixture (e.g., antibodies in an antiserum) a pronounced optimum is usually observed for the dilution. At higher concentrations, too much competition by the other molecules present in the coating sample is experienced, whereas at low concentrations many of the binding sites remain unoccupied. For example, Tijssen *et al.* (1982) observed an optimum dilution of the antiserum at 1:10,000. However, detectability was still significantly lower than the case in which purified IgG was used for coating.

The hydrophobicity of the plastic–protein interaction can be exploited further. In general, proteins have their most hydrophilic residues at the outside, whereas the more hydrophobic residues are oriented toward the inside. Partial denaturation of the immunoreactant exposes these hydrophobic regions and ensures a firmer attachment to the plastic (Ishikawa *et al.*, 1980). This is carried out for IgG as follows (Conradie *et al.*, 1983): purified IgG is dissolved in 50 m*M* glycine–HCl buffer, pH 2.5, containing 0.1 *M* sodium chloride, incubated for 10 min at room temperature, and neutralized with 0.5 *M* Tris. After dialysis against the coating buffer this IgG can be used for sensitization.

For the adsorption of lipid antigens (e.g., mycobacterial glycolipid or cardiolipid), about 2 μg/ml is used in a buffer containing sodium deoxycholate (0.1% w/v), and incubation is carried out for 3 hr at 37°C. The addition of Mg^{2+} to the caoting buffer helps the coating of lipopolysaccharides (Ito *et al.*, 1980).

Covalent linkage of immunoreactants to plastic may be beneficial for many substances. For those for which good results are obtained in the noncovalent attachment, the benefits of covalent attachment are marginal. Glutaraldehyde is suitable for the covalent linkage of those moieties having amino groups. According to the method of Suter (1982), the polystyrene is pretreated with 0.2% (v/v) glutaraldehyde in 0.1 *M* sodium phosphate buffer, pH 5.0, for 4 hr at room temperature. The immunoreactant to be coupled is then added in 0.1 *M* sodium phosphate buffer, pH 8.0, and incubated for 3 hr at 37°C. After washing twice with 0.9% sodium chloride, 0.1 *M* lysine in the pH 8.0 buffer is added and incubated for 1 hr at 37°C. Finally, the plates are washed several times with PBS, containing 0.05% Tween 20. The exact mechanism of this coupling reaction is still unknown. Neurath and Strick (1981) pretreated the plates, before the activation with glutaraldehyde, to generate polyaminopolystyrene by successive incubation with methanesulfonic acid (overnight), extensive washing with water, 4 hr with a mixture of 1:1 glacial acetic acid and fuming nitric acid at 40°C, washing with water until the pH is 6 or higher, and 1 hr at 50°C with 0.5% dithionite in 0.5 *M* sodium hydroxide, and extensive washing with water.

Antigens which bind poorly to plastic may be attached to the solid phase by bridging molecules. Proteins are adsorbed to the plastic, followed by an incubation with an ethanol–acetone (4:1) mixture (30 min at room temperature) followed by three washings with distilled water and air-drying. The antigen may then be conjugated to the solid-phase proteins by any of the methods discussed by Skurrie and Gilbert (1983).

The forms of the plastic supports may vary from microtiter plates (flat bottom, round bottom), well strips, cuvette rows, which are convenient for automation, to bead disks, cocktail stirrers, etc. (Shekarchi *et al.*, 1982). Many of the original tests were based on the use of polypropylene tubes instead of microtiter plates. With tubes it is possible to use a rotating system so that a relatively larger surface is used for the same volume.

Polystyrene is slightly less efficient in retaining the adsorbed material than polyvinyl but yields considerably less background staining.

2. Nitrocellulose Solid Phase

Nitrocellulose membranes (e.g., Transblot from Bio-Rad; 0.45 μm) have long been used in nucleic acid hybridization experiments. This solid phase also adsorbs protein very efficiently and has been used for "dot-immunobinding assays" (Hawkes *et al.*, 1982) and protein blotting of proteins separated by electrophoresis (so-called Western blotting; Towbin *et al.*, 1979).

Minute amounts of sample are applied as droplets on the membrane (0.1–1 μl; spot diameter less than 1 mm). Since adsorption is very effective, this method has a special edge over plastics for the detection of antigens contained in mixtures. Hawkes *et al.* (1982) found that the potential range of antigens to which this technique may be applied is very wide. They tested and obtained excellent results with soluble proteins, nucleic acids, membranes, various organelles, fungi, protozoa, bacteria, and viruses.

Nitrocellulose membranes are rather breakable when dry, but can be easily handled in a wet state. With a pencil, grids (3 × 3 mm) may be drawn on the nitrocellulose sheets or, alternatively, small discs (6 mm) may be punched from them. If several dots will be applied on the same disk, an indexing notch should be cut in each disk for easy location (Walsh *et al.*, 1984).

Before use, the filters are washed with distilled water for 5 min and dried at room temperature. The dotting is achieved by adding 1 μl, containing 100–400 pg (mixtures) or less immunoreactant, to the membrane. The filter should be dried thoroughly to stabilize the binding. Nucleic acid binding requires baking for 60 min at 80°C. At this stage the membranes may be stored for several weeks without any loss of activity. Before detection, the membranes are incubated with PBS containing 0.05% Tween 20 for 2 hr at 37°C to block all binding sites remaining free on the membrane. If these membrane disks, squares, or strips are incubated in the wells of microtiter plates, it is necessary to block the binding sites equally on this plastic.

It is expected that this technique will find applications in many areas. For example, the dotting of different allergens on the same disk allows allergen testing with a very simple method by retrieving IgE antibodies, possibly present, from 50 μl serum samples. For this purpose, it is convenient to use two types of disks, one in which each spot contains, e.g., a pollen mix, a fungus mix, dander mix, and mite mix, and the following disk contains a set of separate dots for each of the allergens of the positive mixture.

Another application of this very promising technique is the rapid and sensitive colorimetric visualization of biotin-labeled DNA probes hybridized to DNA or RNA immobilized on nitrocellulose (Leary *et al,* 1983). Biotin-labeled DNA probes can be prepared by nick translation (Rigby *et al.,* 1977) in the presence of biotinylated analogs of the triphosphate nucleotides and are very useful for the detection of gene sequences. Leary *et al.* (1983) observed that this probe exhibits lower nonspecific binding to nitrocellulose than does radiolabeled DNA. Target sequences in the order of 1–10 pg range can be detected. The DNA dot-blots are prepared by serial dilution of the DNA in 50 m*M* Tris–HCl, pH 7.5/0.3 *M* NaOH. The appropriate dilutions are neutralized on ice with 3 *M* HCl, and 5-μl aliquots are spotted directly on BA-85 nitrocellulose (Schleicher and Schuell). The filters are air-dried and baked for 4 hr at 80°C. This method has the advantage over conventional methods using radioactive probes and autoradiographic detection of prolonged stability of the probes (1–2 years), rapid detection (1–2 hr), and superior resolution. Detectability sufficient for analyzing unique sequences in a 7.5 μg sample of mammalian DNA can be achieved.

Palfree and Elliott (1982) investigated the binding of detergent-solubilized membrane glycoproteins. They observed that Triton X-100 or Tween 80 should be avoided at concentrations higher than 0.01%, whereas deoxycholate, taurocholate, or octylglucoside can be used at higher concentrations. These authors also added a glutaraldehyde fixation step.

3. Paper Solid Phases

Paper can be used as a solid phase after covalent coupling alters activation of paper disks with cyanogen bromide (Giallongo *et al.*, 1982) or with 1-(3-nitrobenzyloxymethyl)-1-pyridinium chloride (Alwine *et al.*, 1979). These methods have not found wide application in EIA.

It is also possible to perform noncovalent dot-immunobinding on paper, at least for water-insoluble proteins (Esen *et al.*, 1983). Its main advantage is its low cost, but it has a lower detectability than similar methods with nitrocellulose. Protein, solubilized in 60% ethanol, is spotted on Whatman No. 1 paper (1–2 μl; 1–2 ng of antigen). The spots are dry after a few minutes and can be stored for EIA.

4. Particulate Solid Phases

Agarose, cellulose, and Sephacryl solid phases have the advantage that they can be held in solution easily, so that the immunoreactants attached can react rapidly with their counterparts in the sample, and have a high capacity for binding due to their large surfaces. The immunoreactants are immobilized on the solid phase by the methods used for the preparation of immunosorbents (Ferrua *et al.*, 1979; Wrights and Hunter, 1982).

C. OPERATIONAL DETAILS OF ENZYME IMMUNOASSAY PROCEDURES

The immune reactions and enzymatic activity are treated separately in this section since many combinations of these may be made. Two general approaches can be used for the immunological stage depending on whether separation steps are required, i.e., solid-phase and homogeneous assays.

1. Enzyme Immunoassays on Solid Phase

The receptor of the molecule to be detected is immobilized on the solid phase as detailed in Chapter 4, Section B. Not only antigens or antibodies may act as receptors but also other molecules with particular affinities such as lectins, protein A, or complement factors.

To these sorbents the molecules or complexes to be detected are added in a suitable buffer (to prevent nonspecific adsorption but to enhance specific interaction), which are in turn detected by enzyme-labeled tracers. Nonspecific adsorption is avoided by adding a large excess of a competing agent for the (nonspecific) binding sites on the solid phase. Generally, this is achieved by the addition of a nonionic detergent (Tween 20, Triton X-100, at 0.05%) or an inert protein (gelatin, BSA at high concentrations). Moreover, the ionic conditions should be such that specific antibody–antigen interactions are not affected or even promoted whereas nonspecific protein interactions are avoided. Usually an isotonic buffer is chosen for this purpose, but to prevent nonspecific interactions the sodium chloride concentration can be raised to 0.3 *M*, since ionic interactions are the most frequent cause of background staining. Another consideration in choosing the buffer is its influence on the enzyme. For example, PBS contains a high concentration of inorganic phosphate which is a strong competitive inhibitor of alkaline phosphatase. For this enzyme, a Tris buffer is, therefore, advised. The buffers which are frequently used are (i) PBS-T (pH 7.4) containing 0.2 g KH_2PO_4, 2.9 g Na_2HPO_4, 8.0 g NaCl, 0.2 g KCl, and 0.5 g Tween 20/liter; (ii) TBS-T (pH 7.4) containing 1.4 g Tris, 8 g NaCl, 0.2 g KCl, 0.5 g Tween 20/liter and adjusted with 1.0 *N* HCl to the correct pH; and (iii) PBS-GM (pH 7.0; advised for β-galactosidase), containing 0.84 g NaH_2PO_4, 0.85 g Na_2HPO_4, 17.5 g NaCl, 1 m*M* $MgCl_2$, 1 g BSA, and 5 g gelatin/liter. Kato *et al.* (1980) obtained better results with the latter buffer after digestion of the gelatine with a protease.

The washing steps between the various incubations define to a large degree the specificity of the test: (i) cross-reacting immunoreactants with low affinity can be selectively removed, and (ii) the undesired protein–protein interactions ("protein-stacking") can be minimized. In cases in which these factors interfere, the number of washing steps can be increased from 3 to 6 (1–5 min each). The volume used in microtiter plates is mostly 0.2 ml, though some investigators (e.g., Yolken *et al.*, 1980) prefer 0.1 ml.

The optimal temperature should be established in preliminary tests. Though the use of 37°C is widespread, it may not be optimal. In fact, this temperature is often chosen in serology for the acceleration of secondary reactions. However, secondary reactions are not considered in EIA, and some antibodies (so-called "cold antibodies") react better (or only) at lower temperatures.

Checkerboard titrations are necessary to determine the optimum dilution of the various components of the test. For the conjugate incubation, two different approaches can be taken. First, after the conjugate has been allowed to react with its antigen adsorbed to the solid phase, it can be used at a concentration giving an absorbance of 1.0 after 30 min of incubation with substrate Second, a dilution can be chosen giving maximum absorbance (i.e., in added in excess) with the specific antigen, but an absorbance below 0.2 is necessary for the nonspecific samples. The choice depends on the method of reading the test. If a microplate reader is used, the first approach is chosen most often since the spectrophotometers are most sensitive and reliable between 0.2 and 1.0. It is evident that the activity of the conjugate should be monitored frequently to detect possible inactivation. For this optimization serial dilutions of conjugate in the appropriate buffer are prepared, and are incubated at 37°C or room temperature for 2 hr. After washing, the substrate (Chapter 4, Section C,3) is added and the absorbance is recorded after 30 min.

Similar tests can be performed to assess the ideal incubation conditions for the other steps. All incubations should be done in humid chambers.

Noncompetitive Solid-Phase Enzyme Immunoassays

The design of these assays has been discussed in Chapter 4, Section A. For the detection of antibodies, antigen is immobilized followed by an incubation with the test sample and the conjugate (enzyme–anti-IgG antibodies, or enzyme–protein A if the IgG in the test sample is reactive with protein A).

This test is performed by first establishing the optimum coating conditions. Serial dilutions of the antigen are made and, subsequently, the solid phase is sensitized as described in Chapter 4, Section B. After washing, serially diluted reference sera (positive, weakly positive, and negative) are added and incubated for 2 hr at room temperature. Following another washing step, the conjugate is added at the optimal concentration (determined as described above), incubated for 3 hr at room temperature, washed again, and followed by the addition of substrate.

The indirect method has about 10-times higher detectabilities than the direct method. Even higher detectabilities are possible with the bridge methods, though specificity tends to decrease. Three types of bridging methods are currently used: (i) the avidin–biotin method, (ii) the PAP method, and (iii) the lectin bridge method.

The avidin–biotin method is based on the extraordinary affinity of avidin for biotin. Biotin can be coupled relatively easily to all kinds of molecules. Generally, *p*-nitrophenyl esters of biotin or *N*-hydroxysuccimide esters of biotin are added (0.1 *M* in dimethylformamide or dimethyl sulfoxide) to the antibody or enzyme solutions (at 20 mg/ml) at a ratio of 100 μl for IgG, 60 μl for glucose oxidase, 40 μl for β-galactosidase (enzyme activity somewhat affected), and 5 μl for peroxidase (not recommended for alkaline phosphatase; Guesdon *et al.*, 1979). This mixture is incubated for several hours and then dialyzed extensively. A frequent test sequence is solid-phase antigen, antibody from test sample, biotinylated enzyme. This sequence can be obtained, however, by three different methods: enzyme-labeled avidin–biotin (LAB; Guesdon *et al.*, 1979), the bridged avidin–biotin (BRAB; Guesdon *et al.*, 1979), and the avidin–biotin complex (ABC; Hsu *et al.*, 1981).

The LAB is simple, is less time consuming than other avidin–biotin methods, and has reasonable detectability. Moreover, it can be applied to systems with enzymes which are inactivated by biotinylation (such as alkaline phosphatase). In this method, after incubation with the test sample and washing, 200 μl of biotinylated antibody (at 0.5–1.0 μg/ml) is added, washed after incubation for 2 hr, and 200 μl of 0.25–0.1 μg/ml solution of enzyme-conjugated avidin is added and incubated for 2 hr at room temperature.

In the BRAB method, after the incubation with the biotinylated antibody and washing, avidin is added (200 μl/well; at 10 μg/ml of avidin) and incubated for 20 min (pronounced optimum) at room temperature. After four washings, with the buffer used for dilution, 200 μl of a 1.0 μg/ml solution of biotinylated enzyme is added and incubated for 1 hr at room temperature.

The ABC is prepared prior to its addition to the biotinylated anti-IgG antibodies immobilized in the wells. For this purpose a 10 μg/ml solution of avidin (in PBS-T) is added to an equal volume of a 4 μg/ml solution of biotinylated peroxidase (also in PBS-T). This is left for at least 15 min at room temperature (to allow the formation of the ABC complex), and is then added to the wells and incubated for 15 min at room temperature.

The ABC and BRAB methods have higher detectabilities than the LAB method, with the exception of alkaline phosphatase-mediated tests, due to the sensitivity of this enzyme to biotinylation. Washing should be extensive. Increasing the ionic strength will probably be beneficial in that it decreases nonspecific ionic interactions of the very basic avidin with immobilized proteins.

Immunologically linked antibodies have been popular in enzyme immunohistochemistry for the past 15 years (Sternberger, 1979) and may become popular for EIA as well, particularly with the advent of monoclonal antibodies. With the latter, soluble, specific antibody–enzyme complexes are obtained simply by admixture since these antibodies will recognize only one epitope on the enzymes and thus be unable to form complexes with more than one antibody. If antibody immobilized by the antigen on the solid phase is obtained from the same species as the anti-enzyme antibody, it may be linked with an anti-IgG (Butler *et al.*, 1980). Since several enzyme complexes may thus be linked to a single antigen–antibody complex, detectability increases significantly. In some cases, it may be possible to replace the linking antibody by protein A (Yolken *et al.*, 1980). In detail, in the procedure using immunologically linked enzyme–antibody complexes, optimum conditions of coating are established as described above. Subsequently the primary antibody (positive, weakly positive, negative) is serially diluted and incubated for 2 hr at room temperature (one vertical row/dilution). After washing, twofold serial dilutions (in horizontal rows) of the bridging antibody are incubated for 2 hr at room temperature to establish the concentration at which the required excess is obtained. An excess of enzyme–antibody complexes (e.g., 25 μg/ml) is added and incubated for 2 hr at room temperature. Optimum conditions can be determined after the enzyme activity is known.

Lectins can also serve as bridging molecules (Guesdon and Avrameas, 1980) if the affinity of the lectin is sufficient. For example, antibody may be conjugated with Con A, followed by an incubation with peroxidase.

The detection of antigens requires the immobilization of antibody on the solid phase and results in so-called sandwich assays. Different designs are possible (Chapter 4, Section A), of which the most common is the direct sandwich assay in which the antibody is immobilized on the solid phase and serves to extract the antigen from the sample, followed by detection using enzyme-labeled antibodies. Generally this approach has a detectability of about 10 ng/ml. In the so-called double-sandwich methods, the antigen is detected by the indirect method. Since the enzyme-labeled anti-IgG antibody should not react with the antibody coated on the solid phase, the latter should be obtained from a species which does not cross-react with anti-IgG antibody (Crook and Payne, 1980). Alternatively, Fab, immobilized on the solid phase, will not react with anti-IgG antibody or with protein A (Barbara and Clark, 1982), since these are reactive almost exclusively with the Fc fragment.

The incubation periods and conditions for these tests are similar to those described above for the tests in which the antigen is immobilized on the solid phase. However, if complete antiserum is used for the sensitization of the plates a dilution of 10,000 is most appropriate (Tijssen *et al.*, 1982).

An interesting side effect is that with the coating of Fab a narrowing of the specificity of the antiserum is observed (Koenig and Paul, 1982).

Other immunosorbents may be prepared by coating C1q on the solid phase. This complement factor selectively extracts antibody–antigen complexes (Calcott and Müller-Eberhard, 1973; Yolken and Stopa, 1980). For this purpose, a 2 μg/ml solution of C1q in the carbonate buffer is used for the coating. Serial dilutions of the antigen are added to each well with antibody, incubated for 3 hr at room temperature, washed, and detected with labeled anti-IgG or anti-enzyme antibodies.

In the class-capture methods (Duermeyer and van der Veen, 1978) antibodies to a certain immunoglobulin class are immobilized and serve to extract these immunoglobulins from the serum sample. If among these immunoglobulins antibodies are present, the subsequently added antigen and labeled anti-antigen antibodies will be bound as well. A comparison of the relative titers obtained with tests for different classes reveals the distribution of the antibodies in these classes. Rheumatoid factors may, however, interfere in IgM class-capture assays (Fig. 4.8).

2. Enzyme Immunoassays in Fluid Phase

Most of the assays in this group are of the competitive variety and are used for haptens. Haptenated enzymes are incubated with antiserum. The antibody reacting with the hapten on the enzyme modifies the enzyme activity. The presence of the hapten in the sample decreases this modulation since it will react with the antibody and thus prevent its action on the enzyme. Most of these assay kits are marketed under the trade name EMIT (Syva Co.). The enzymes used in these assays are different from the solid-phase assays (e.g., lysozyme, glucose-6-phosphate dehydrogenase, malate dehydrogenase). A problem associated with this approach is the possible interference by substances in the samples with the enzyme action or immune reactivity since there are no separation steps. Moreover, some antisera may contain antibodies that decrease enzyme activity and antibodies that increase enzyme activity (Rowley *et al.*, 1975).

An example is the original method of Rubenstein *et al.* (1972) for the detection of morphine. Carboxymethyl morphine is labeled to lysozyme by means of the mixed anhydride method. The sample (50 μl urine) is added to 200 μl substrate (*Micrococcus luteus*) at 37°C. Sequentially, 50 μl of hapten antibody and enzyme-labeled hapten is added. The enzyme-catalyzed reaction is measured over a 40-sec period (decrease of light scattering at 450 nm). Prolongation of the period of measurement increases detectability about 30 times.

Similarly, substrate-labeled hapten (Burd *et al.*, 1977) and antibody–enzyme conjugates (Wei and Reibe, 1977) can be used.

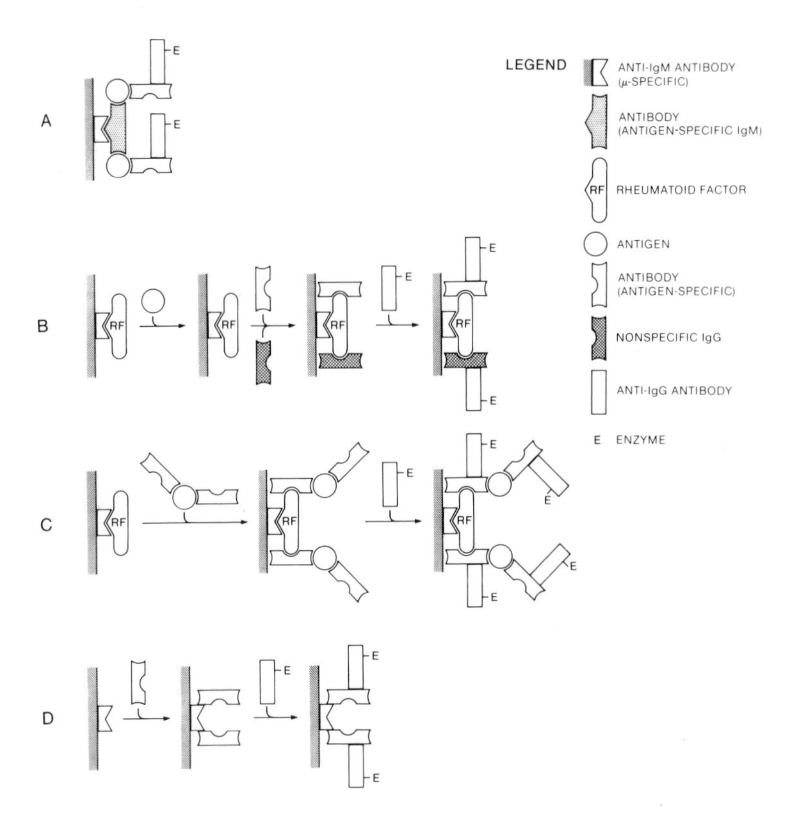

Fig. 4.8. Causes of nonspecificity in indirect class-capture assays. Diagram A demonstrates the complex that should be formed to obtain specific results (Fig. 4.3). Rheumatoid factor (RF), an IgM, will capture specific and nonspecific IgG which will be detected by the conjugate (B), or capture immune complexes (C). Finally, nonspecific protein–protein interactions may cause false positive results (D).

Recently, an enzyme-linked immunosorbent assay was developed for detection of antigen/antibody reactions in liquid-phase (McCullough *et al.*, 1985). This liquid-phase ELISA procedure is used to assay both antisera and hybridoma cultures and to demonstrate the greatest discrimination of antibodies species. Both quantitatively and qualitatively fluid-phase ELISA is at least as efficient as the double sandwich or antigen-trapping ELISA, but eight times more sensitive than the indirect ELISA. This procedure provides an efficient possibility for the detection of immunologically relevant epitopes.

Competitive Solid-Phase Enzyme Immunoassays

The original enzyme immunoassay (van Weemen and Schuurs, 1971) is based on this principle.

In the case of immobilized antibody, two approaches can be used to determine the antigen concentration, i.e., simultaneous or sequential competition. In the first, the labeled and test sample antigens, which compete for the same antibody on the solid phase, are added simultaneously; in the second, the test antigen precedes the labeled standard antigen. The presence of antigen will, in both cases, decrease the activity.

An example of a sequential competitive assay is as follows. The solid phase is coated with antibodies (Chapter 4, Section B). Then a serial dilution of labeled antigen is added to establish a convenient concentration of the latter (should not be in excess; Chapter 4, Section A). In the test, dilutions of a solution with a known concentration of antigen is added to the antibody-coated wells, incubated for 2 hr at room temperature, and followed, after washing, with the enzyme-labeled antigen to saturate those binding sites still free. Incubation conditions are as described in Chapter 4, Section C,a with the adjustment of the concentration of antibody to that of optimum inhibition as determined by checkerboard titrations.

The immobilization of antigen on the solid phase allows the competition of antigen with the solid-phase antigen for the same enzyme-labeled antibody. The sequential approach of this method has, e.g., been applied for the detection of common sequences on proteins (Tijssen and Kurstak, 1981). Altschuh and van Regenmortel (1982) applied this method for an investigation of the antigenic composition of tobacco mosaic virus, whereas Friguet *et al.* (1983) reported the use of this method for the distinction of specificities of monoclonal antibodies.

3. The Measurement of Enzyme Activity

Most quantitative assays are based on the build-up of the colored product which is measured, e.g., after 30 min.

Peroxidase is widely used. It should be realized that the substrate, hydrogen peroxide, is also a powerful inhibitor so that best results are obtained at a defined concentration (Kurstak *et al.*, 1977; Tijssen *et al.*, 1982). The reduction of peroxide by the enzyme is achieved by hydrogen donors which can often be measured conveniently after oxidation.

The choice of the chromogenic donor often depends on preferences of the laboratory since many donors may be satisfactory. In some methods, such as enzyme immunohistochemistry, immunodotting, or "Western" blots, the soluble chromogenic donor should precipitate as a colored product on oxidation. In other cases the converted substrate should remain soluble to prevent any scattering in absorptiometric methods such as with microplate readers. Diaminobenzidine is used chiefly in enzyme immunohistochemistry, both at light and electron microscopic levels. With nitrocellulose, however, 4-chloro-1-naphthol is preferred. Several comparative studies indicate that the most satisfactory candidates for the optical density methods are *o*-phenylenediamine (OPD), 5-aminosalicylic acid (5-AS), *o*-toluidine (OT), 2,2′-azinodi(ethylbenzothiazolinesulfonic acid) (ABTS), and MBTH/DMAB [3-methyl-2-benzothiazolinone hydrazone and 3-(dimethylamino)benzoic acid] (Saunders *et al.*, 1977; Voller *et al.*, 1979; Tijssen *et al.*, 1982; Al-Kaissy and Mostratos, 1983).

ABTS is prepared as a solution of 40 mg donor per 100 ml of a 0.1 *M* phosphate-citrate buffer, pH 4.0. A similar concentration is generally used for OPD, but in a 0.1 *M* sodium citrate buffer, pH 5.0. However, recent investigations have shown that higher concentrations of substrate are beneficial (2 mg/ml; Gallati and Brodbeck, 1982). The pH of the substrate should be adjusted to 5 after the donor has been added. OPD should be prepared just before use and should be shielded from light. *o*-Dianisidine (10 mg/ml in methanol) is diluted just before use by adding 0.9 ml to 100 ml 0.1 *M* phosphate-citrate, pH 5.0, but is in our experience not as useful as the other donors.

MBTH/DMAB (Tijssen *et al.*, 1982; Geoghegan *et al.*, 1983) is prepared from different stock solutions: 30 m*M* DMAB is dissolved in 0.1 *M* dibasic sodium phosphate, adjusted to pH 7.0 with citric acid, and stored in lightly closed amber bottles at 4°C. A stock solution of 0.6 m*M* MBTH is dissolved in a similar buffer and stored in the same conditions just before use; 7 volumes of buffer, with 1 volume of MBTH and 1 volume of DMAB are mixed.

Commercial 5-AS is dissolved in 100 ml of distilled water at 70°C for about 5 min with stirring. After cooling to room temperature, the pH of the solution is raised to 6.0 with a few drops of 1 *M* sodium hydroxide.

The optimum substrate (hydrogen peroxide) concentration depends both on the H-donor and the solid phase which affects the microenvironment of the enzyme. This can be established by simple preliminary tests. Generally concentrations between 0.009 and 0.0006% are adequate. The optimum is pronounced, due to the substrate inhibition of the enzyme at higher concentrations. The addition of 0.05% Tween 20 to prevent inactivation of peroxidase was recommended by Porstmann *et al.* (1981).

The development of the colored product is measured at different wavelengths. The optimum wavelength may also shift if the reaction has been arrested by adding a blocking reagent to prevent a change of the optical density after the reaction period. These blocking reagents are 0.5 volume hydrochloric acid for OPD, 0.1 volume of 1 *M* sulfuric acid for DMAB/MBTH, and 0.5 volume of 10% sodium dodecyl sulfate for ABTS and 5-AS. The optimum wavelengths are 415 nm for ABTS, 492 nm for acidified OPD (445 for nonacidified OPD), 590 nm for both nonacidified and acidified DMAB/MBTH, and 492 nm for 5-AS (Kurstak *et al.,* 1984).

For the detection of peroxidase on nitrocellulose blots, 4-chloro-1-naphthol is prepared. About 30–40 mg of chloronaphthol is dissolved in 0.2–0.5 ml absolute alcohol. This is added to 50 m*M* Tris buffer, pH 7.6, containing 50 μl hydrogen peroxide while stirring, heated to 45°C and filtered (generally a white precipitate arises) while hot.

β-Galactosidase activity is determined after adding a solution containing 70 mg *o*-nitrophenyl-β-D-galactopyranoside per 100 ml 0.1 *M* potassium phosphate buffer, pH 7.0, containing 1 m*M* magnesium chloride and 0.01 *M* 2-mercaptoethanol. The generation of nitrophenol can be followed at 405 nm. The reaction may be stopped by adding 0.25 volume of 2 *M* sodium carbonate.

Alkaline phosphatase should be assayed in a buffer depending on the origin of the enzyme. For the bacterial enzyme 0.1 *M* Tris–HCl buffer, pH 8.1, containing 0.01% magnesium chloride is used, whereas for the intestinal mucosal enzyme a 10% (w/w) diethanolamine (97 ml in 1 liter of a 0.01% magnesium chloride solution) buffer, pH 9.8 (adjusted with hydrochloric acid) is used. Just before use *p*-nitrophenylphosphate is added to these buffers at 1 mg/ml. The production of nitrophenol is monitored at 405 nm. The reaction is stopped by adding 0.1 volume of 10 m*M* cysteine.

Alkaline phosphatase is also useful for histochemical detection of antigens on nitrocellulose blots. Particularly useful are the reagents described by Blake *et al.* (1984) and Knecht and Dimond (1984). The stock solutions made were (i) 5-bromo-4-chloroindolyl phosphate (5 mg/ml in dimethylformamide), (ii) nitro blue tetrazolium (1 mg/ml in 0.15 *M* veronal acetate, pH 9.6), and (iii) 2 *M* magnesium chloride. Just before use, 20 μl of magnesium chloride, 1 ml of the nitro blue tetrazolium, 0.1 ml of the indolylphosphate, and 9 ml of veronal acetate buffer are mixed. Knecht and Dimond (1984) observed that a lower pH (8.8, with 0.1 *M* Tris–HCl) produced less background staining. It is not clear from their paper whether the bacterial enzyme was used. They described an elegant method for obtaining a transparent duplicate of the blot. For this purpose, they prepared a solution of 0.1 *M* Tris–HCl, pH 8.3, and 1 m*M* magnesium chloride, containing 1% agarose, heating to dissolve the latter and cooling to 65°C, then the substrate (dissolved in a small volume of dimethyl sulfoxide) was added. This solution is poured on a glass plate, and the dry, washed nitrocellulose blot (with the protein side down) is placed on the gel. When the color is sufficiently developed, both the blot and agarose (between two cellophane sheets) are dried separately. Wetting enhanced faded color to its original intensity.

Glucose oxidase is most conveniently detected by adding peroxidase in excess (25 μg/ml) together with substrate and H-donor for the latter.

Fluorigenic substrates may increase detectability of the assay considerably. This necessitates filtering of the water, since extraneous fluorescent material is often present, and the use of spectrofluorometers (though UV light boxes may also be used (Forghani *et al.*, 1980).

For peroxidase, 3-hydroxyphenylpropionic acid (HPPA) is most appropriate (Zaitu and Okhura, 1980). This yields a product which shows the greatest degree of fluorescence at high pH. HPPA is added to 0.1 *M* sodium phosphate buffer, pH 8.0, at a rate of 5 g/liter. The pH is thereby decreasing to 7. This substrate is added to the solid-phase peroxidase, and after 5 min ⅕ volume of 0.03% hydrogen peroxide is added. After the desired incubation period (10–100 min), an equal volume of 0.2 *M* glycine–sodium hydroxide (pH 10.3) is added. The excitation wavelength is 320 nm (emission at 405 nm).

The enzyme giving highest detectability with fluorigenic substrates is β-galactosidase. The hydrolysis of 4-methylumbelliferyl-D-galactopyranoside yields fluorescent 4-methylumbelliferone. For this assay a 10 m*M* sodium phosphate buffer, pH 7.0, containing 0.1 *M* sodium chloride, 1 m*M* magnesium chloride, 0.1% sodium azide, and 0.1% BSA is prepared. To 100 ml of this buffer, 0.34 ml of a 1% solution of 4-methylumbelliferyl-D-galactopyranoside in *N,N'*-dimethylformamide is added. The "stop-buffer" consists of 2.0 *M* glycine–sodium hydroxide, pH 10.3. The excitation wavelength is 360 nm and the emission wavelength is 440–450 nm.

Alkaline phosphatase is assayed similarly. The only difference is the substrate for which 4-methylumbelliferyl-phosphate is used. Compared to the β-galactosidase substrate this substrate has the disadvantage that background levels, due to spontaneous hydrolysis, are considerably higher (Neurath and Strick, 1981).

In homogeneous enzyme immunoassays, separation steps are not necessary, and the influence of a solid phase on the enzyme activity is absent.

These enzymes are therefore assayed under the standard conditions usually employed for these enzymes.

D. INTERPRETATION OF RESULTS OBTAINED WITH ENZYME IMMUNOASSAYS

Enzyme immunoassays are performed to establish either the antibody activity or the antigen concentration. The quantification or activity determination of these two immunoreactants is quite different. The antigen is generally well defined and concentrations in milligrams per milliliter may be obtained. However, antibodies are very heterogeneous with respect to affinity and physicochemical properties. Moreover, antibodies are generally assayed over a wider concentration span than the antigens.

An important observation for antibody dose–response curves is that these curves are rarely parallel due to differences in concentration and affinity. Therefore, at one dilution one serum may indicate a higher activity, whereas at another dilution (at the other side of the cross-over), the inverse result would be obtained. Another problem is the occurrence of antibodies in different antibody classes which could cause a prozone effect (de Savigny and Voller, 1980). Moreover, the affinity of the antibodies trapped by the immobilized antigen tends to change with the dilution. Lehtonen and Eerola (1982) observed that at low dilutions antibodies of high affinity are preferentially bound, whereas at high dilutions antibodies with a wider range of affinities are bound. This problem has not yet been solved satisfactorily.

These differences should be taken into account for the different modes of expression of the results. It is also necessary to stress the problems associated with the determination of the discrimination level to distinguish between positive and negative results. Intra- and interassay variations of results need to be assessed. To improve these parameters, different methods have been adopted to linearize the sigmoidal dose–response curves, and computer programs have been designed to rapidly process EIA data.

1. The Cutoff Value

Irrespective of the method chosen, at some stage a decision has to be made as to whether a response is indicative of the presence or absence of antibodies. Most methods have a built-in discrimination level. This level may be set at 0.15 or 0.20 absorbance (Halbert *et al.*, 1983), two or three times the mean (Malvano *et al.*, 1982), or the mean plus two or three standard deviations (SD) (Richardson *et al.*, 1983). It should be stressed that the estimate of SD is subject to error as well, and depends on the number of responses from which it was calculated (Kurstak, 1985).

A compromise generally has to be made for the setting of the cutoff level since irrespective of its value, false positives (type I errors) or false negatives (type II error) are encountered. Moving the level to higher values will minimize false positives at the cost of more false negatives and vice versa. It is also possible to define responses in this intermediate area as doubtful (Heck *et al.*, 1980), e.g., between the mean of negatives and the mean plus four standard deviations.

General experimental approaches have been proposed to reduce the expenditure and time involved in establishing the cutoff value. Van Loon and van der Veen (1980) established the mean and standard deviation for a large group of normal sera, and, in parallel, tested reference sera. The mean plus three standard deviations of the negatives equalled 40% of the absorbance of a certain reference serum, which was included in subsequent tests as an internal standard for the cutoff value. Cremer *et al.* (1982) included both standardized positive and negative sera.

2. Methods of Expressing Serological Results in Enzyme Immunoassays

a. *Semiquantitative Tests*

Visual inspection has frequently been used for rapid screening of enzyme immunoassays on microtiter plates or on other solid phases, such as nitrocellulose (immunodotting). The detectability is of the same order of magnitude as absorptiometric methods. However, it is quite subjective and difficult to distinguish positive from negative responses at the doubtful interval. Moreover, such titration methods yield a discontinuous scale of results and require a series of dilutions. Nevertheless, the titer obtained is proportional to antibody activity.

A more precise end point is obtained by determining the optical density at each dilution and establishing the dilution at which the dose–response curve intersects the cutoff level.

b. *Effective Dose Method*

The effective dose method, proposed by Leinikki and Passila (1977), compares the dose–response curves obtained by a dilution series of the test

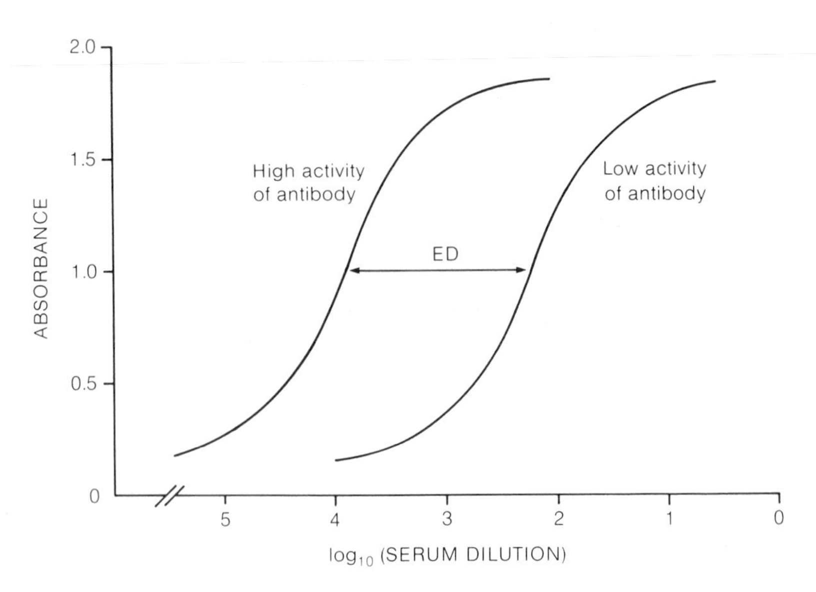

Fig. 4.9. The determination of the effective dose (ED). The optical density in wells with increasing concentrations of the serum to be tested and of a reference serum is compared at the highest sensitivity (i.e., at the steepest section of the dose–response curve). The difference in dilution of the sera to obtain this same optical density is expressed in Brigg's logs.

serum and a dilution series of the positive reference serum. The difference between the dilutions at the steepest point of the dose–response curves of two sera is expressed in Brigg's logs (Fig. 4.9).

Reproducibility of this method is better than that of the titration method since in the latter the result is established in an area of low precision (tail of the sigmoid curve). The effective dose is also proportional to antibody activity. Disadvantages of the effective dose method are the time-consuming and expensive manner of activity measurement (similar to the titration method) and the uncommon mode of expressing the effective dose (e.g., effective dose = −0.49).

c. The Absorbance Method

The use of chromogenic substrates leads almost automatically to the use of absorbance methods. Few or only one suitable dilution needs to be made to obtain the result. However, there is no direct relation of antibody activity to the optical density, and the latter is not proportional to antibody activity. de Savigny and Voller (1980) stressed the limited value of absorbance values to clinicians.

An advantage, in addition to the single-dilution requirement, is that the scale is continuous, though it may be less reliable outside the 0.2–0.8 units range.

d. Standard Unit Curves

In this method the absorbance values are transformed to standard units, which yield a continuous scale and are proportional to the antibody activity. Moreover, single dilutions of each serum are needed if it is assumed that dose–response curves for the various sera are parallel. This method is illustrated in Fig. 4.10.

The standard curve is constructed from plotting the absorbance obtained

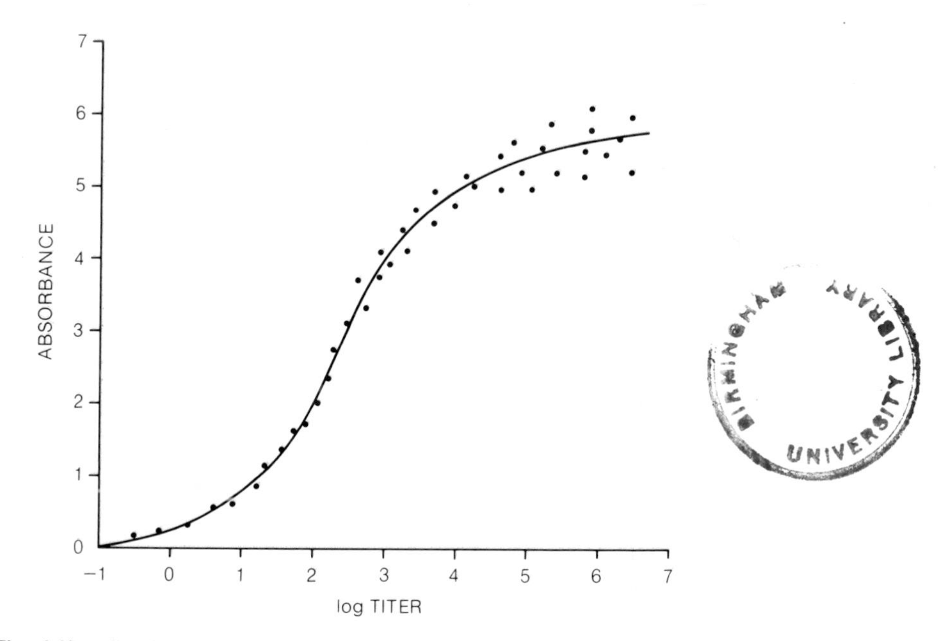

Fig. 4.10. Standard curve obtained from a large number of sera (negative, weakly positive, strongly positive) by plotting the absorbance at a fixed dilution (e.g., 1:1000) against the titer of that serum. Variation increases significantly with the titer. The sigmoid curve may be linearized by several methods, making it easier to estimate the titer of serum if a certain optical density is obtained. The standard units obtained are directly proportional to antibody activity.

from sera at a single dilution (e.g., 1:1000) against their respective titers. The test samples in subsequent experiments are measured for activity at the same dilution (internal standards from the original curve can be included) and the titer is obtained directly from the standard curve. Malvano *et al.* (1982) used a similar approach, except that instead of a battery of different sera, one positive serum, calibrated against a WHO standard serum, diluted to varying degrees with a negative serum was used to construct the standard-unit curve.

With these approaches it is necessary to include internal standards for between-run, between-laboratory, and between-method normalization.

e. Ratio Methods

Different ratio methods are used in which the test sample absorbance is expressed in terms of a reference serum.

The P/N (positive/negative serum absorbance; Locarnini *et al.*, 1979) is frequently employed. A P/N greater than 2 or 3 is considered positive. This method is simple and easily understood, and neither time consuming nor expensive. However, reproducibility is poor (ratio dependent on particular negative reference serum) and ratios are not directly proportional to antibody activity.

Sedgwick *et al.* (1983) devised a ratio method based on the areas under the dose–response curves (at a few dilutions). This method does not assume that dose–response curves are parallel. The ratios obtained are proportional to antibody activity and have better reproducibility. It is, however, also more time consuming and expensive.

f. Multiple of Normal Activity (MONA)

Felgner (1978) considered the lower half of the sigmoidal dose–response curve and calculated the parabola exponent n from this part (Fig. 4.11) with the formula

$$\frac{(A_{ser})^{N}}{A_{ser.\ dil}} = \text{dilution factor}$$

in which A_{ser} is the absorbance at fixed dilution and $A_{ser.\ dil}$ the absorbance at further dilutions. This A_{ser} should be below the point of inflection on the dose–response curve.

The absorbance of the test serum can be expressed in MONA by the following formula:

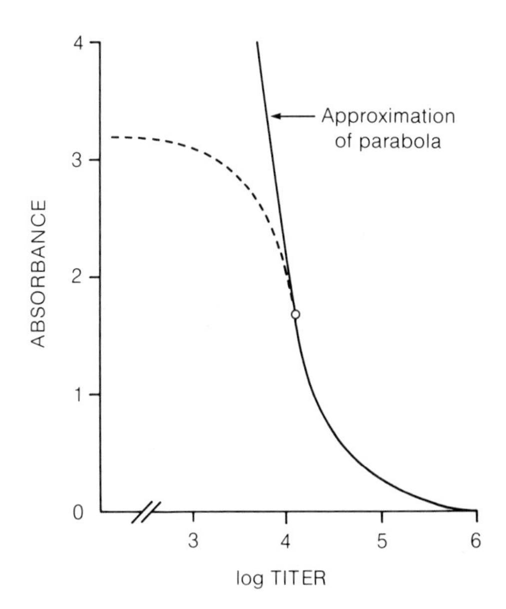

Fig. 4.11. The determination of MONA values according to Felgner (1978). The sigmoid curve is considered until its midpoint (indicated by ○), and the lower part of this dose–response curve can be approximated by a parabola, by calculation of the parabola exponent taking several serum dilutions. In this approach the leveling off of the dose–response curve at high antibody concentrations is artificially avoided.

$$\log_{10} \text{MONA}_{\text{test}} = n(\log_{10} A_{\text{test}} + \log_{10} A_{\text{neg. ser}})$$

The parabola exponent may change from experiment to experiment, but the MONA will be rather stable.

This method has several advantages such as (i) single dilution and improved sensitivity, (ii) proportionality to antibody activity, and (iii) comparison of the result directly to the negative serum.

It is assumed in this approach that dose–response curves are parallel and the determination of n may be greatly influenced by this nonparallelism.

3. Linearization of Dose–Response Curves

Linearization is particularly popular in hapten concentration determinations. Several mathematical models have been developed (see Oellerich *et al.*, 1982):

Four-parameter log–logit model:

$$R = R_0 + K_c\left\{\frac{1}{1 + \exp[-(a + b \ln C)]}\right\}$$

Five-parameter logit:

$$R = R_0 + K_c\left\{\frac{1}{1 + \exp[-(a + b \ln C + cC]}\right\}$$

Five-parameter exponential:

$$R = R_0 + K \exp[a \ln C + b(\ln C)^2 + c(\ln C)^3]$$

Five-parameter polynomial:

$$\ln C = a + b\left[\frac{(R - R_0)}{100}\right] + c\left[\frac{(R - R_0)^2}{100}\right] + d\left[\frac{(R - R_0)^3}{100}\right]$$

Spline approximation:

$$c = a_i + b_i (R - R_i) + c_i (R - R_i)^2 + d_i (R - R_i)^3$$

Where R is the rate of change of absorbance (R_0 the predicted rate at zero concentration), K_c the predicted difference between the maximum and minimum response (at infinite and zero dose, respectively), K a scale parameter for five-parameter exponential model, C the concentration of the standards, and *a, b, c,* and *d* various parameters to account for nonlinearity.

Oellerich *et al.* (1982) concluded that the four-parameter log–logit model is most practical. The logit is defined (Berkson, 1944) as $\ln(y/1-y)$ where y is a measurable response value.

The four-parameter log–logit model is analogous to the Hill equation (Atkinson, 1966; Rodgers, 1984) in enzyme kinetics, though some thermodynamic restrictions should be made (De Lean *et al.*, 1978). In commercial kits, the K_c given may quite often have to be changed for the particular laboratory conditions (Dietzler *et al.*, 1980) otherwise curves may not be straight. The slope is quite often low (about 0.5–0.8; Rodbard and McClean, 1977) so that a wide range of concentrations can be detected, but results in low sensitivity, Linearization generally leads to an artificial compression of errors at either end of the curve.

5

Enzyme Immunohistochemical Methods

A. INTRODUCTION

Since its inception by Coons *et al.* (1950) immunofluorescence has been used to great advantage in immunodiagnosis for the localization of antigens *in situ* and also for titration purposes. Titration by immunofluorescence has been replaced almost completely by EIA, whereas for localization studies the fluorescein label is increasingly replaced by enzymes. This approach makes it possible to eliminate some of the problems and deficiencies of the immunofluorescence technique. The enzyme converts a soluble substrate into a highly insoluble product which precipitates at the site of the enzyme. These precipitates should be visible (light microscopy) or should be able to be rendered visible or electron opaque (e.g., osmification of oxidized diaminobenzidine makes it electron opaque), thus enabling ultrastructural localization. In summary, enzymes instead of fluorescein labels have the following advantages: they can be used at both light and electron microscopic levels, reagents are stable, nonlabeled cellular components which can be rendered visible by simple methods, only ordinary microscopes are required, for most enzymes the preparations are very stable, and designs for detection can be much more sophisticated, resulting in higher detectability.

The fluorescein label was initially replaced by peroxidase (Avrameas and Uriel, 1966; Nakane and Pierce, 1966), and the technique has been commonly known as immunoperoxidase (Kurstak *et al.*, 1969). Gradually more enzymes were found suitable for specific purposes (e.g., for double immunoenzymatic labeling; Mason and Woolston, 1982). The immunoperoxidase technique was adopted for virology in our laboratory about 15 years ago (Kurstak *et al.*, 1969; Kurstak, 1971), and its success is attested to by the numerous publications on this subject (reviewed by Sternberger *et al.*, 1970; Sternberger, 1979; Kurstak, 1971; Kurstak and Kurstak, 1974; Kurstak *et al.*, 1971, 1972, 1975, 1977, 1984; Nakane and Farr, 1981; Bullock and Petrusz, 1982; Polak and van Noorden, 1983).

Fixation of the tissue is probably the most important step for reliable localization without affecting the immunogenicity of the antigens, to prevent diffusion of the antigens (more frequent than generally assumed), to permeabilize the membranes, and to make the antigens accessible without changing the gross conformation of the tissue. Thus, a compromise should be made between complete and rapid fixation, on the one hand, and permeabilization and preservation of immunoreactivity, on the other. Not surprisingly, improper fixation is the main obstacle in demonstrating intracellular antigens (Bohn, 1978).

B. FIXATION OF TISSUE

Detailed descriptions of fixation procedures are found in Culling (1974), Nairn (1976), and Pearse (1980). The antigens to be detected can be soluble or incorporated into structural components of the cell. Fixation should arrest enzyme action to prevent structural decomposition, hinder diffusion of soluble components, and fortify the tissue.

Generally, two types of fixatives can be distinguished: the precipitants, which are most suitable for immobilized antigens, and the cross-linkers (e.g., formaldehyde), which are most suitable for soluble antigens. Diffusible antigens can be distinguished from bound antigens by the study of cryostat sections. Prefixation diffusion artifacts (e.g., passive uptake of nonspecific antigens by cells; Mason *et al.*, 1980) can be counteracted by rapid fixation. Nevertheless, postfixation artifacts are also frequent, e.g., with ethanol. It is self-evident that small antigens are best preserved by cross-linkers.

A problem frequently encountered with cross-linking fixatives is that the epitopes of interest may be masked, resulting in false negative results. For example, Brandtzaeg (1982) noted that formalin masks antigenicity rather than denaturing the antigen. Diffusion artifacts, on the other hand, lead to false positive results (incorrect localization). The signal-to-noise ratio can be improved by the extraction of diffusible antigens in cold saline.

Organic solvents, such as methanol and acetone, are excellent precipitant fixatives, though some antigens may be removed and lipids extracted. Commonly used are 100% methanol, 100% acetone, or 5% acetic acid in absolute ethanol. Slides or coverslips carrying cell preparations are submerged for 10 min in one of these solutions at −20°C. After fixation it is necessary to wash the preparation in cold saline to attain neutrality.

Numerous formulations of cross-linking fixatives have been described. In particular formaldehyde is popular, whereas glutaradlehyde (a dialdehyde) yields increased cross-linking, but tends also to decrease antigenicity.

The common commercial formaldehyde solution (37% formalin) is not suitable, since it contains a number of stabilizers (Farr and Nakane, 1981). Formaldehyde can be obtained by depolymerizing paraformaldehyde. For this purpose, 10 g of paraformaldehyde is dissolved in 50 ml distilled water under continuous stirring. A few drops of added 0.1 *M* sodium hydroxide will clear the solution. This solution is then diluted with the appropriate buffer (e.g., 0.11 *M* sodium phosphate buffer, pH 7.5) to 50 ml and 1 g of sodium chloride is added (end concentrations: 2% formaldehyde, 0.1 *M* buffer, and 0.2% sodium chloride).

Buffered formaldehyde solutions often yield inconsistent fixation and antigen preservation (Mason *et al.*, 1980). In combination with glutaraldehyde, they may yield superior results, particularly for relatively small antigens such as lysozyme (Rodning *et al.*, 1980). However, for larger antigens increasingly lower glutaraldehyde concentrations (1% and less) should be used. Baker's formol calcium, a mild cross-linking fixative, is particularly useful for phospholipids (Ranki *et al.*, 1980).

Sometimes an addition of precipitant fixatives to the formaldehyde solution is beneficial. Stefanini *et al.* (1967) added picric acid. This solution is prepared by adding and depolymerizing paraformaldehyde (13.3%) in double-filtered saturated aqueous picric acid solutions, and diluting 6.6 times with buffer. This fixative resembles Bouin's fixative, which has, however, a higher formaldehyde and acetic acid concentration.

Mercuric chloride is also a powerful protein precipitant and can be combined with formaldehyde (formol sublimate, pH 3), trichloroacetic acid, and acetic acid [Heidenhain-Susa (*Sub*limat *Säu*re, sublimate acid) or potassium sublimate and acetic acid (Zenker's fluid)]. Among these, formol sublimate was found to give superior results (Bosman *et al.*, 1977; Brandtzaeg, 1982).

Overfixation by formaldehyde can frequently be reversed by protease treatment. Antigens concealed during fixation can be unmasked by protease treatment to restore antigenic reactivity (Huang, 1975; Radaszkiewicz *et al.*, 1979; Stein *et al.*, 1980). Matthews (1981) and Isobe *et al.* (1972) reported that proteolytic digestion may also eliminate xylol-induced antigen impairment and, to a lesser degree, glutaraldehyde overfixation. Pronase at 1 g/liter (for 15 min) in 0.05 *M* Tris–HCl, pH 7.8, or trypsin (in 0.01% calcium chloride, pH 7.8) for 1 hr at 0.5 g/ml is the most commonly used enzyme. It should be realized that the balance between over- and underdigestion is fine.

Benzoquinone is a weak bifunctional cross-linking agent with tissue constituents and can be used at, e.g., 0.5% in phosphate-buffered saline for 30 min at room temperature. This reagent has been successful for the vizualization of neuropeptides or small molecules (Jessen, 1983).

Nonionic detergents, such as Triton X-100 and NP-40, are often used for the cytoskeleton. They are used at 0.3% in PIPES buffer, containing 0.1 m*M* EGTA, for about 2 min. Increased permeabilization by repeated freeze-thaw cycles in the presence of antiserum to immobilize the antigens as rapidly as possible can sometimes be helpful. The coverslip covered with serum is placed on a metal plate cooled with dry ice and thawed.

Epitopes in antigens may normally be hidden (e.g., J-chain in dimeric IgA) and as such are undetectable in tissues. Brandtzaeg (1976) unmasked such epitopes by a treatment for 1 hr at 4°C in 0.1 *M* glycine–HCl, pH 3.2, containing 6 *M* urea.

C. DETECTION OF ANTIGENS AT THE LIGHT MICROSCOPIC LEVEL

After fixation, preparations should be properly washed to bring the pH to neutrality in order to induce antibody–antigen interactions and to prevent a deleterious effect on antibodies or enzyme labels. Most isotonic buffers at pH 7.0 are suitable.

Incubations are performed in humid chambers. However, caution is warranted for the temperature, since some antibodies (''cold antibodies'') react poorly at 37°C but much better at lower temperatures.

1. Procedures with Antibodies with Covalently Linked Enzymes

Three types of procedures are most often used, corresponding to those in EIA discussed in Chapter IV, Section A,1. These are the direct, indirect, and anticomplement methods, which have increasing detectabilities in this order (Kurstak *et al.*, 1978) and are represented in Fig. 5.1.

The direct method is simple and needs few controls. However, for each antigen to be detected a separate labeled antiserum is required. Antibodies are incubated with the preparation at a concentration of about 50 μg/ml for about 1 hr. Before the histochemical staining they should be well rinsed (e.g., 3 times for 5 min in PBS).

In the direct method, labeled anti-IgG antibodies or labeled protein A is used to detect primary antibodies completed by the antigens *in situ*. This method thus requires two incubation steps, but a single conjugate can be used for different antisera and different antigens. Moreover, detectability increases at least five fold. The primary antibodies need not to be purified. The concentrations at which the primary antibodies and the conjugate are used are about 10 and 30 μg/ml, respectively.

The anticomplement method is a variant of the indirect method in which complement is fixed by the immune complexes, and subsequently detected by labeled anticomplement antibodies. The complement can be added directly to the primary antiserum (or can be present in fresh primary antiserum). This method has several advantages over the indirect method, such as higher detectability and the fact that a single conjugate can be used for complement-fixing antibodies from different mammalian species.

Indirect methods carry the disadvantage that nonspecific antibodies bound to the tissue will also be detected with great ease resulting in higher background levels. Complement is fixed in the neighborhood of the immune complex (Mardiney *et al.*, 1968), and the precise location may be more difficult to determine particularly with electron microscopy.

2. Detection with Immunological Labeled Enzyme

Since enzymes are antigenic, this property may be used to introduce enzymes at the desired location. For example, a tissue antigen is allowed to react with its antibody from species A, followed by an excess of antibody (from species B) to immunoglobulin from species A. This is followed by

DIRECT METHOD

1.- DETECTION OF ANTIGEN

+ DAB + H_2O_2

2.- DETECTION OF AUTOIMMUNE COMPLEX

INDIRECT METHOD

1.- DETECTION OF ANTIGEN

+ DAB + H_2O_2

2.- DETECTION OF AUTOIMMUNE COMPLEX

+ DAB + H_2O_2

3.- DETECTION OF COMPLEMENT FIXING ANTIBODY-ANTIGEN COMPLEX

+ DAB + H_2O_2

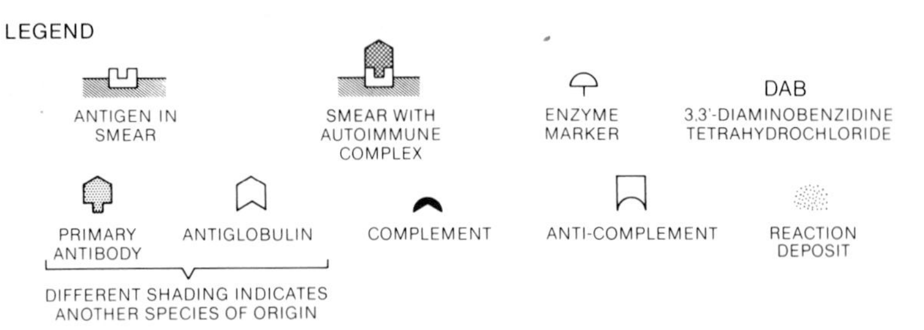

Fig. 5.1. Direct and indirect immunoperoxidase methods. The direct method is convenient if large quantities of antigen and antibody are available since it is not very sensitive and a conjugate has to be prepared for each antigen. The direct method is, however, convenient for the detection of autoimmune complexes. The indirect method can be extended to anti-complement immunoperoxidase procedures if the antibodies fix complement.

an incubation with anti-enzyme antibodies, raised in species A, which will be bound to the anti-immunoglobulin. The enzyme added later will then be trapped by the anti-enzyme antibodies (Fig. 5.2; Mason *et al.*, 1969; Sternberger, 1969; Dougherty *et al.*, 1974).

This approach has important limitations. The immunoglobulin from the anti-enzyme antiserum trapped by antibodies from species B need not and will not be specific for the enzyme. Usually only a fraction of less than 10% is specific for the enzyme. Therefore, many nonspecific immunoglobulins will be bound, thus increasing the detectability.

A powerful improvement was introduced by Sternberger *et al.* (1970) in which the anti-enzyme antiserum is mixed with enzyme resulting in the precipitation of specific enzyme–antibody complexes. These are purified and resolubilized by simultaneously adding enzyme and lowering the pH (Sternberger *et al.*, 1970). These soluble peroxidase–antiperoxidase (PAP) complexes are then added in one step, instead of the two last steps (anti-enzyme and enzyme incubations) in the original method (Fig. 5.3.). PAP complexes are rather large, which may hamper their penetration into cells, particularly for electron microscopic studies.

Recently, the availability of monoclonal anti-enzyme antibodies resulted in considerably smaller complexes since the monoclonal antibodies recognize only one epitope on the enzyme. (It should be realized that the originally large differences possible in the detectabilities among chemically linked antibodies have now shrunk to insignificant levels. However, this approach may be used to further increase detectabilities, i.e., double-bridge, etc., by adding (Vacca *et al.*, 1978), after the normal sequence, other anti-immunoglobulin and enzyme–antibody complex incubation steps (Fig. 5.3).

The primary antibody is diluted about 500 times and (incubate for 1 hr at room temperature) with the preparation washed several times with isotonic buffer and incubated with a 1:25 dilution of the anti-immunoglobulin antiserum. After washing, the anti-immunoglobulin is added at about 40 μg/ml. Pickel *et al.* (1976) observed that the use of 3% sodium chloride in the buffer considerably decreased background staining. Triton X-100 or Tween-20 at low concentration (0.05%) in the primary antiserum step was also reported beneficial.

Instead of immunologically linked enzyme, avidin–biotin complexes can be used (Hsu *et al.*, 1981). Biotin-labeled anti-immunoglobulin is used at 1:200 dilutions, followed by an incubation with a solution containing 8 μg/ml avidin and 2 μg/ml biotin peroxidase.

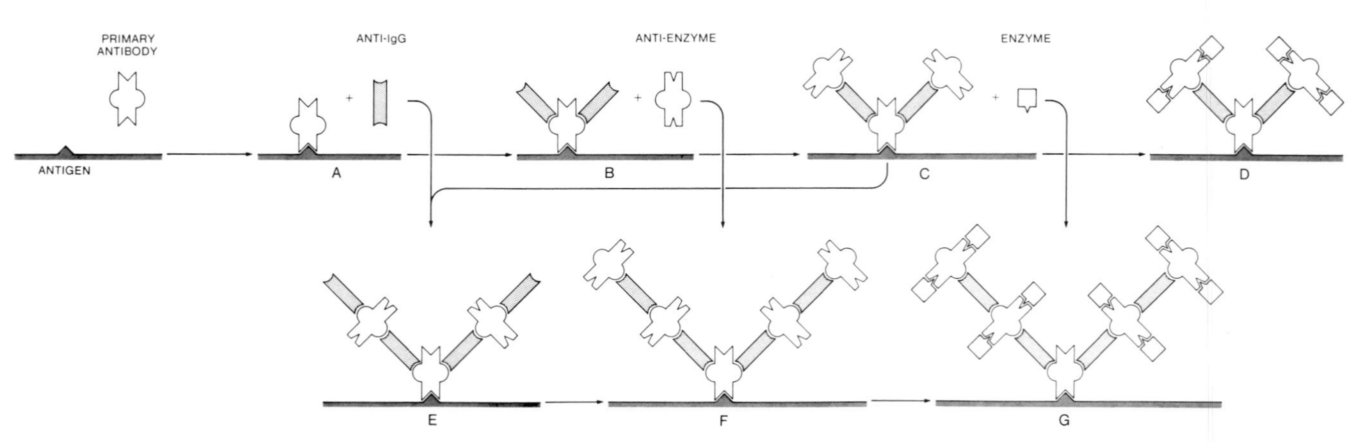

Fig. 5.2. Double-bridge method to enhance detectability. The primary antibody reacts with the antigen (A, B) and is, in turn, detected by several anti-IgG antibodies (added in excess). Due to the excess most of the second IgG binding Fab parts of the antiimmunoglobulin will remain free (C) and can immobilize subsequently added anti-enzyme antibody if the latter was prepared in the same animal as the primary antibody (D). At this point enzyme can be added (single bridge), or the sequence of adding anti-immunoglobulin and anti-enzyme be repeated (F, G). The enzyme is then added. Clearly, a large amplification is obtained by this double bridge. It should be realized, however, that this increase in detectability is often at the cost of specificity. It is also important that the anti-enzyme antibody is pure (e.g., monoclonal antibody) since the binding of an antibody with another specificity will decrease the detectability of this test.

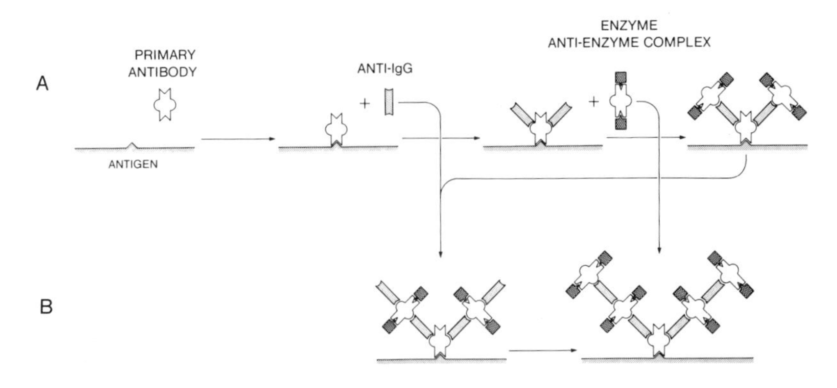

Fig. 5.3. Single-bridge (A) and double-bridge (B) enzyme–anti-enzyme complex method. The primary antibody produced in species I is allowed to react with the antigen in the smear. An anti-IgG immunoglobulin produced in species II is added in excess so that only one of the paratopes of the antibodies reacts with the immobilized antibody. A soluble enzyme–anti-enzyme antibody complex prepared from antibody produced in species I and enzyme is added and will be immobilized by the immunoglobulin on the site of the antigen. This sequence can be repeated (B) to obtain a double bridge and thus enhanced staining.

D. DOUBLE IMMUNOENZYMATIC LABELING

Hybridoma techniques provide an increasing availability of antibodies against single epitopes. Therefore, there will be an increasing demand for techniques that allow the discriminate detection of epitopes in the same cell (Mason and Sammons, 1978; 1979). For this purpose, it is necessary to have enzyme reactions that can be recognized distinctly. For light microscopy, the color of the products should be different, whereas for electron microscopy a different opacity for the two complexes is required.

In some situations it is possible to use the same enzyme for both components of the double immunoenzymatic labeling, if this enzyme will yield products of contrasting colors with the different substrates. However, it is then necessary to localize these epitopes sequentially. With different enzymes, the immunoreactants can be added simultaneously, minimizing the time required. An example of the latter is the use of peroxidase with diaminobenzidine/hydrogen peroxide and alkaline phosphatase with naphthol AS-MX/Fast Blue BB which give a brown and blue product, respectively. Endogenous alkaline phosphatase is blocked with levamisole (1 m*M* added to the substrate solution; Ponder and Wilkinson, 1981).

It is essential in these procedures that components of the sandwiches used for the detection of the two epitopes not interreact. Prior to the staining procedure (Mason and Woolston, 1982), preparations are equilibrated with neutral buffer and with 1:10 diluted normal sheep serum for 10 min. This will reduce nonspecific staining. The preparation is then incubated for 30 min with the primary antibodies against the two epitopes at suitable concentrations (1–10 μg/ml). These primary antibodies (e.g., mouse monoclonal and rabbit polyclonal antibodies) should not cross-react with an anti-immunoglobulin antibody. After washing, the antisera, prepared in the same species (e.g., sheep) against the two primary antibodies, is added in excess (dilution 1:25), followed by a washing and addition of anti-enzyme complexes (e.g., rabbit PAP, and monoclonal mouse anti-alkaline phosphatase). After washing, the peroxidase appears after the addition of diaminobenzidine tetrahydrochloride at 0.6 mg/ml and 0.01% hydrogen peroxide in Tris-buffered saline. The reaction can be followed by microscopic observation and is stopped before background staining develops (2–7 min). Alkaline phosphatase appears after the addition of naphthol-AS-MX dissolved in dimethylformamide (2 mg/0.2 ml) in a glass tube. Tris (0.1*M*) is added for a final volume of 10 ml. Immediately before staining 1 mg Fast Blue BB is added. The slides washed in Tris-buffered saline are shaken dry and the substrate is added (if cloudy it should be filtered onto the slide). After color development (10–15 min) the slides are washed. They should be mounted in aqueous material because alcohol or xylol dissolves the alkaline phosphatase reaction product.

Another approach is the sequential application of the sandwiches. The antibodies need not be obtained from different species, but a complete elution of the antibodies is required from the first sandwich before applying the second sandwich (Vandesande, 1979). Sternberger and Joseph (1979) avoided this elution step by using peroxidase and diaminobenzidine in the first step. Diaminobenzidine apparently masked the antigen and the first sandwich. For the second step, 4-chloro-1-naphthol can be used. This substrate is prepared by dissolving 40 mg 4-chloro-1-naphthol in 0.2 ml ethanol, which is then added to 100 ml of 50 m*M* Tris–HCl buffer, pH 7.6, containing 0.003% hydrogen peroxide. The precipitate is removed by filtration. Since the diaminobenzidine color product is dominant, no mixing of colors at the original sandwich site is observed (brown; chloronaphthol yields a blue product).

E. NONSPECIFIC STAINING

Nonspecific staining may be due to endogenous enzymes, immunological nonspecificity, nonspecific adsorption of sera, or methodological errors.

Endogenous peroxidase may be blocked by pretreatment of the preparations with 0.5% hydrogen peroxide in methanol for 15 min, whereas endogenous alkaline phosphatase is selectively and effectively blocked by including levamisole at 1 m*M* in the substrate solution. The intestinal alkaline phosphatase used for labeling is not affected by this inhibitor.

Immunological nonspecificity is widespread and can even be used to great advantage. For example, Marucci and Dougherty (1975) used guinea pig antibodies to human immunoglobulins to link human antibodies to PAP complexes prepared from baboon antibodies. Control and specificity tests (Fig. 5.4) and blocking and neutralizing methods (Fig. 5.5) can be performed to establish the absence of undesired staining.

Undesired immunological nonspecificity requires a change in methodology or the removal of cross-reactivity by affinity or adsorption methods. Nonspecific adsorption can be eliminated to a large extent by pretreating the preparations with normal sera, by raising the ionic strength of the incubation media two- or threefold (Pickel *et al.*, 1976), or by the inclusion of Triton X-100 or Tween-20 at 0.05% as generally used in serological enzyme immunoassays. Another cause of nonspecific uptake of immunocytochemical reagents is the reaction of these reagents with fixative in the preparation. Fixation with cross-linkers may make the preparation sticky, e.g., for free amino groups in the conjugate (Molin *et al.*, 1978). To avoid this problem sections may be treated with 0.01 *M* lysine prior to immunoenzymatic labeling.

F. ULTRASTRUCTURAL LOCALIZATION OF ANTIGENS

Two approaches may be distinguished, the ''preembedding'' and the ''postembedding'' technique, depending on whether the immunological steps are performed prior to or after embedding. Electron microscopy requires an excellent ultrastructure preservation, but for the localization of the antigens, antigenicity must be maintained. The best-suited fixatives and fixation method thus depend largely on the properties of the antigen. Preembedding also requires a permeabilization of the tissue to increase accessibility of the antigens for the relatively large immuno complexes. This permeabilization should not be derimental to the morphology of the cell or increase diffusion artifacts. On the other hand, in postembedding staining, in which the tissue is fixed, embedded, and sectioned prior to staining, difficulties may arise: the embedding material may mask or destroy epitopes, epitopes only at the surface of the section may be detected, and these may be nonspecific uptake of the immunocytochemical reagents

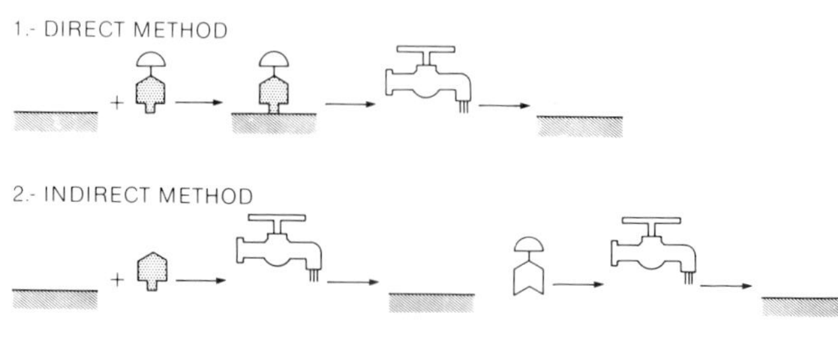

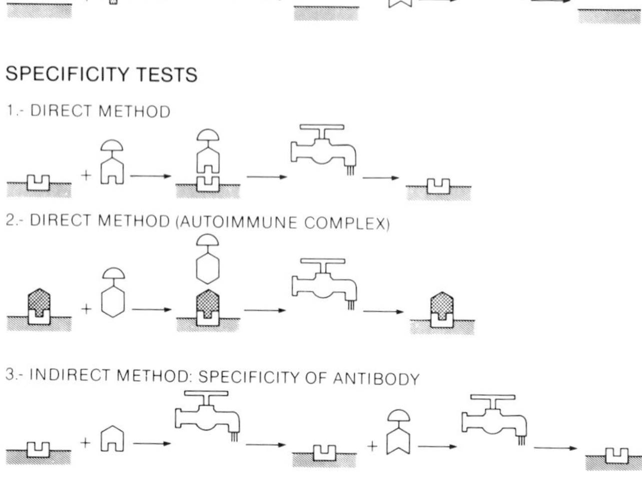

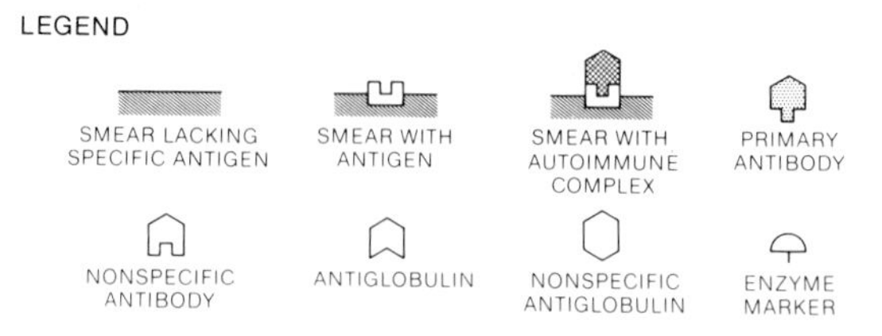

Fig. 5.4. Control and specificity tests for antigen in smear and for antibody. The direct and indirect tests for antigen in the smear should be negative if the smear lacks the antigen (control tests) or if the specific conjugate is replaced by a nonspecific conjugate. The specificity test can be extended to determine the specificity of the anti-globulin.

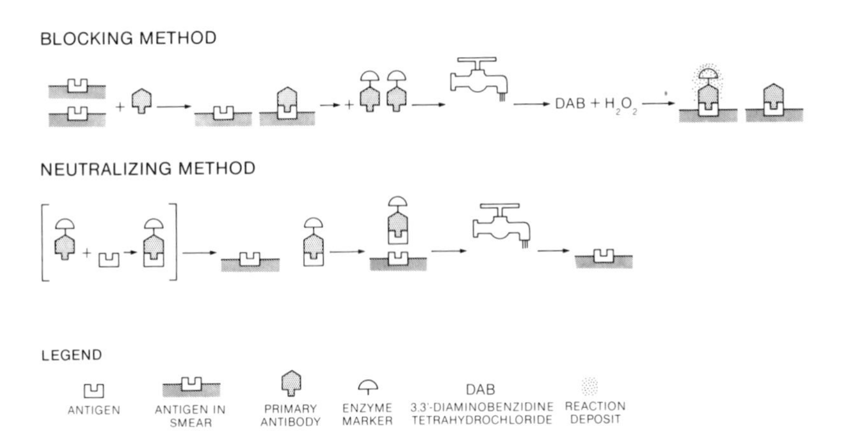

Fig. 5.5. Controls for the immunoperoxidase method using specific antigen or antibody. Two different approaches can be taken to ascertain the specificity of the reaction. The blocking reaction is used when specific antibody is available. This control antibody will decrease the staining of specific peroxidase-conjugated antibody but not of nonspecific peroxidase-conjugated antibody. The neutralizing method is used when specific antigen is available. Prior incubation of the soluble specific antigen with the conjugate will decrease the staining, whereas nonspecific antigen will not influence the amount staining.

by the embedding plastic, similar to that which occurs in solid-phase enzyme immunoassays.

The enzyme generally used in these techniques is peroxidase since the conversion product of diaminobenzidine becomes electron opaque when treated with osmium tetroxide. For preembedding purposes, microperoxidase conjugated to Fab fragments may offer a significant advantage with respect to size but is somewhat less reactive (Tiggemann *et al.*, 1981). It should be mentioned that copper grids are not suitable for the immunoperoxidase method since they react with osmium tetroxide; nickel or gold grids should be used instead. Fixatives commonly used are Bouin's fluid, glutaraldehyde (low concentration, less than 0.5%), paraformaldehyde mixtures, periodate–lysine–paraformaldehyde (PLP method, MacLean and Nakane, 1974), or Zamboni's picric acid formaldehyde.

1. Preembedding Technique

The preembedding technique consists of a preliminary fixation (prefixation), antigen–antibody reaction, postfixation, enzyme reaction, embedding, sectioning, and staining. Osmium tetroxide, in general, sharply decreases antigenicity and is not recommended as a prefixative. The method of Bohn (1980) is illustrative of this approach. Cell monolayers are rinsed with 0.2 *M* phosphate buffer (pH 7.3) and treated with a mixture of glutaraldehyde (0.125%), formaldehyde (1%), and saponin (0.5 mg/ml) in the same buffer for 5 min at 4°C. The cells are then fixed for 45 min in the same buffer without saponin (which permeabilizes the membranes), rinsed for 30 min with buffer at 4°C, and incubated with 2% albumin in the buffer for 1 hr (to reduce the stickiness of the tissue, i.e., unblocked cross-linkers used for fixation). Tissue fragments can, e.g., be prefixed by the PLP method which is based on a limited oxidation of sugars and subsequent cross-linking.

The prefixed cells are well washed and subjected to immunocytochemical reactions. It has been found advantageous to incubate with high dilutions of primary sera for prolonged periods (e.g., 2 days). Washing should also be performed over extended periods since antibodies may enter or leave the cells quite slowly. Postfixation is generally done with glutaraldehyde since peroxidase and microperoxidase are not very sensitive to this cross-linker. A fixative of 2% glutaraldehyde in PBS–10% sucrose is suitable and can be added to the cells for 30 min at 4°C to prevent further deterioration of the tissue. Sometimes this is followed by an incubation with a 1% solution of an inert protein in PBS–sucrose for 10 min, air-drying, and another cycle of postfixation. The peroxidase appears as described above for light microscopy but for longer periods (e.g., 10 min; heavy staining required). The enzyme product is made electron opaque by incubation with 2% osmium tetroxide for 30 min. The specimen is then dehydrated, embedded, and sectioned (glass knives) by the usual methods.

2. Postembedding Techniques

Postembedding procedures (Ordronneau, 1982) maintain structural detail most faithfully but the requirement of antigen preservation remains. The embedding material may be of some importance, for example, the widely used epoxy embedding media are very reactive and may mask epitopes. Methacrylates do not copolymerize with cellular constituents and may better preserve the cellular integrity. PAP and bridge-methods (Fig. 5.4) are very popular in this approach. The plastic may cause nonspecific uptake of immunoreactants which can be counteracted by incubating with normal serum and low concentration of nonionic detergents. The disadvantage of PAP and double-bridge methods is that the complexes are large and the staining products may spread over the surface near the epitope. In some instances it is possible to include an etching step to unmask the epitopes from the plastic. Instead of the 0.01% benzene and 1% methanol for methacrylate and 1 and 10%, respectively, for Epon, Moriarty and Holmi (1972) used hydrogen peroxide, since this reagent seems to swell the plastic thereby improving accessibility.

Typically, a minced tissue is fixed overnight in 4% paraformaldehyde and 0.1% glutaraldehyde in 0.01 *M* phosphate buffer, pH 7.4. After washing in PBS the tissue is embedded, and sections are cut (about 75 nm; light gold to silver) and placed on nickel or gold grids (300 mesh). Sections (Araldite or Epon) are etched for 10 min on a few chops of 10% hydrogen peroxide (not necessary for methacrylate). The sections are incubated with normal serum (1:20) in an isotonic buffer. After immunoreactions and enzyme appearance, grids are rinsed with Tris-buffered saline and PBS, respectively, and treated for 15 min with 2% osmium tetroxide.

Enzyme Immunoassays after Immunoblotting

The high detectability of enzyme immunoassays and the refined methodology of transferring macromolecules to membranes after separation by electrophoresis resulted in a powerful adjunct for biochemical and medical research. In the past 2 years, there has been an enormous increase in the application of these techniques.

A. TECHNICAL APPROACHES

After separation of the proteins in denaturing or nondenaturing gels by electrophoresis or electrofocusing, they can be transferred to membranes by three different methods (Gershoni and Palade, 1983): diffusion, liquid flow transfer, and electrophoretic transfer.

The membrane most often used in immunoblotting is nitrocellulose; the more efficient Zetabind membranes give generally unacceptable background staining (Gershoni and Palade, 1983). Initially nitrocellulose membranes with a porosity of 0.45 μm were used, but small proteins tend to pass through these membranes at high current densities (e.g., Lin and Kasamatsu, 1983) and a smaller pore size (0.1 μm) reduces losses. Conditions of transfer from the gel to the membrane are given in Table 6.1. SDS may pose problems since SDS–protein complexes adhere poorly to mem-

TABLE 6.1
Protein Transfer from Gel to Membranes

Gel	Conditions for nitrocellulose membranes	References
Urea gels	In 0.7% acetic acid; 1hr at 6 V/cm	Towbin *et al.* (1979)
SDS gels	In 25 m*M* Tris–192 m*M* glycine and 20% methanol; 1 hr at 6 V/cm	Towbin *et al.* (1979)
SDS gels	In Tris buffer, pH 7.4–4 *M* urea (diffusion, 40–48 hr)	Bowen *et al.* (1980)
Thin nondenaturing gels (after isoelectric focusing)	10 m*M* Tris–HCl, pH 8 (2 hr diffusion)	Reinhart and Malamud (1982)
SDS gels, isoelectric focusing gels	Vacuum-induced solvent flow (20–60 min)	Peferoen *et al.* (1982)
SDS gels	25 m*M* Tris–192 m*M* glycine plus 0.01% SDS (15 min at 6 V/cm), or with 20% methanol (1 hr at 6 V/cm)	Nielsen *et al.* (1982)
Agarose gels	3 hr diffusion	Lanzillo *et al.* (1983)
Gel and membrane	**Conditions for covalent linkage**	**References**
SDS gels/DBM paper	25 m*M* phosphate buffer, pH 6.5 (10 V/cm for 1 hr)	Bittner *et al.* (1980)
SDS gels/diazophenylthioether paper (in borate/sulfate buffer)	10 m*M* sodium borate pH 9.2 (1 hr, 32 V/cm)	Reiser and Wardale (1981)
SDS gels/DBM paper	Equilibration in phosphate buffer and periodate cleavage, 16 hr	Renart et al. (1979)

branes. In the original method (Towbin *et al.*, 1979) SDS–protein complexes (negatively charged) were electrophoretically transferred to a membrane in a medium without SDS but with methanol. In later reports, low concentrations of SDS in the transfer medium were found to improve transfer. Ericksen *et al.* (1982), Nielsen *et al.* (1982), and Sutton *et al.* (1982) included 0.02 to 0.1% SDS which may prevent the precipitation of proteins in the gel due to a more rapid elution of SDS than of proteins. Transferred proteins can be reversibly stained with heparin and toluidine blue at low pH (Towbin and Gordon, 1984).

Before the proteins transferred to the membrane can be assayed with enzyme immunomethodology (similar to the immunoperoxidase technique), membranes should be treated to prevent nonspecific adsorption of immunoreactants. Inert proteins are often chosen for this purpose. Gelatin, which is frequently used, is often unsatisfactory. Burnette (1981) found whole serum to be superior (not if protein A is to be used in the detection of specific proteins), while Towbin and Gordon (1984) use 10% horse serum. We found the use of Tween 20 as a blocking agent (Batteiger *et al.*, 1982) to be very useful and routinely incubate the blot with 0.05% of this detergent in PBS for 90 min at 37°C directly after the transfer. In all subsequent incubation steps this detergent is added in the same manner. Transferred proteins which were denatured by SDS can be renatured to some extent with nonionic detergents, such as Triton X-100 and Nonidet P-40 (Petit *et al.*, 1982) or zwittergents.

The optimal condition of incubation times and antibody dilutions for the immunodetection depend on various factors which differ for most tests. Antibodies with low affinity will have a very short half-life. Consequently such antibodies should be applied in large excess, but short incubation periods suffice since dissociation is quite rapid (e.g., $k_d = 10^2 \text{ sec}^{-1}$) and equilibrium is reached rapidly. On the other hand, high-affinity antibodies have low dissociation rates (e.g., $k_d = 10^{-2} \text{ sec}^{-1}$) and low concentrations of antibodies are used for detection but for long periods since equilibrium is reached later. For antibodies with average to good affinities, a concentration of 100 ng/ml and an incubation period of 1 hr suffice for maximum complex formation.

After conjugation, the association constant tends to decrease and incubation periods should be prolonged, e.g., by 50%, and the concentration increased about 5–8 times. Among other things, this depends on the size of the enzyme and the degree of aggregation occurring. Peroxidase markers instead of β-galactosidase will have less influence on the diffusion rate and thus lower the association constant less than with β-galactosidase. The enzymes most often used are peroxidase, alkaline phosphatase, or glucose oxidase (Table 6.2). Among the substrates used for peroxidase, 4-chloro-1-naphthol gave the best results. This substrate is, however, not very soluble and should be dissolved in a minimum amount of ethanol and added to Tris-buffered saline; heating to 50°C and filtration while hot increase the staining.

In addition to the application of this technique to diagnosis, it is applicable in numerous areas of research. Antibodies may be eluted from a certain band and assayed on others to establish possible relationship (Olmsted, 1981). Epitope mapping (Clark, 1983) is possible on proteins digested by

TABLE 6.2
Enzymes and Substrates Used in Immunoblotting

Enzyme	Substrate	Antigen detected	References
Peroxidase	*o*-Dianisidine/H_2O_2	Ribosomal proteins	Towbin *et al.* (1979)
	DAB/H_2O_2	Dehistonized chromatin	Glass *et al.* (1981)
	DAB/H_2O_2 and enhancement with nickel and cobalt ions and osmium postfixation	Synaptosomal plasma membranes	DeBlas and Cherwinski (1983)
	3-Amino-2-ethyl carbazol/H_2O_2	Angiotensin-1-converting enzyme	Lanzillo *et al.* (1983)
	4-Chloro-1-naphthol/H_2O_2	Various antigens	Hawkes *et al.* (1982)
Alkaline phosphatase	Naphthol ASMX phosphate/fast Red TR	*Salmonella* proteins	O'Connor and Ashman (1982)
	β-Naphthyl phosphate/fast Blue B	Various antigens	Turner (1983)
	α-Nahthyl phosphate/fast Blue B	IgG	Jalkanen and Jalkanen (1983)
Glucose oxidase	Peroxidase—glucose and hydrogen donor	Viral antigens	Porter and Porter (1984)

proteases and separated from the peptides by two-dimensional electrophoresis (Tijssen and Kurstak, 1983; Kurstak *et al.*, 1984).

An important aspect of immunoblotting, as well as enzyme immunoassays in general, is that interacting molecules other than antibodies can be tested. Some possible reasons for failure are given in Table 6.3.

B. IMMUNOBLOTTING APPLICATIONS IN DIAGNOSIS AND RESEARCH

Nitrocellulose membranes on which antigens are blotted can be dried and stored over prolonged periods. Therefore, they can be used for routine immunodiagnostic tests (reviewed by Towbin and Gordon, 1984). This approach has been used for the diagnosis of infections by worms, protozoa,

TABLE 6.3
Troubleshooting in Immunoblotting

Poor binding to membrane
Membrane is a mixed ester nitrocellulose (pure nitrocellulose should be used)
Proteins are too small, i.e., pass membrane (use 0.1 μm pore size)
Proteins may be removed by SDS, NP-40, or some other detergents (use Tween-20 in washes)
ABM paper is functional if yellow color is not intense
Insufficient sample load
High background levels
Improper blocking (e.g., too low concentration) (hemoglobin reacts with peroxidase; BSA may contain IgG)
Blocker shares common epitopes
Impure second antibody or excessive incubation periods
Insufficient postantibody reaction washes (stronger detergents can be used)
Low enzyme-product formation
Incomplete binding of antigens (see Poor binding to membrane)
Immunologic reactions incomplete (test by immunodotting)
Enzyme of conjugate inactive (enzyme inactivated, e.g., by excessive peroxide concentration)
Substrate solution inadequate
Nonspecific binding
Monoclonal antibodies may react nonspecifically with denatured (e.g., by SDS) proteins
Excessive concentration of antibodies
Presence of cross-reactive IgG

mycoplasma, chlamydia, bacteria, viruses, or for hormones, tumor antigens, allergy, fertility.

1. Infectious Diseases

Immunoblotting proved a very important adjunct for the diagnosis of infectious diseases (Towbin and Gordon, 1984). This procedure is particularly advantageous in cases of poor solubility, where organisms are difficult to cultivate, or where minimum amounts are present. McMichael *et al.* (1981) established that the electrophoretic mobility of HBsAg is related to patient status, and Gupta (1982) investigated this antigen in circulating immune complexes. Such immune complexes may give rise to nonspecific staining in some experiments if rheumatoid factors are present (Chapter IV, Section A,1,a). Immunoblotting is an elegant method for the distinction of profiles of antibodies against the separated, individual polypeptides. For example, Eberle and Mou (1983) used this method to compare profiles to herpes simplex virus (HSV) polypeptides for the antisera from normal individuals or those with persistent lesions. This technique allowed

Hall and Choppin (1981) to demonstrate that patients with subacute sclerosing panencephalitis after measles infection lack one of the viral proteins normally encountered, suggesting the production of defective virus. On the other hand, Dörries and ter Meulen (1983) found, while studying the antibody profile on blotted proteins from Coxsackie virus, that the IgM antibody was directed to a specific polypeptide. This method proved very useful for the identification of insect and plant viral antigens in field isolates (Rybicki and von Wechmar, 1982; Roseto *et al.*, 1982; Hohmann and Faulkner, 1983; Knell *et al.*, 1983).

The electrophoretic separation of proteins prior to immunologic testing may enhance the specificity. For instance, Newhall *et al.* (1982) and Saikku *et al.* (1983) analyzed the immune response to individual polypeptides generated by chlamydia. Another example illustrating this important advantage is the study of Hanff *et al.* (1983) who searched by means of immunoblots the polypeptides common or unique for pathogenic and nonpathogenic strains of *Treponema*. Baughn *et al.* (1983) could, by means of this technique, identify the antigens from *Treponema* present in immune complexes.

Immunoblotting appears very useful for the investigation and diagnosis of protozoan infections such as the malarial *Plasmodium* (Anders et al., 1983; Cappel *et al.*, 1983; Freeman and Holder, 1983) and *Toxoplasma* (Partonen *et al.*, 1983; Sharma *et al.*, 1983).

2. Autoimmunity

The first application of immunoblotting (Towbin *et al.*, 1979) was with ribosomal epitopes. Immunoblotting allowed further characterization of the relevant antigens in connective tissue diseases (Gordon and Rosenthal, 1984). Cross-reactivity with bacterial antigens may cause autoimmune syndromes (Cunningham and Russell, 1983; Zabriskie and Friedman, 1983). Anti-Epstein–Barr virus nuclear antigen (EBNA) antibodies are frequently associated with rheumatoid arthritis (Billings *et al.*, 1983), whereas some histones were shown by immunoblotting to be important autoimmune antigens in systemic lupus erythematosus (Hardin and Thomas, 1984).

3. Tumor Antigens

EBNA, which can be shown by immunoperoxidase procedures to be present in transformed lymphocytes such as Raji cells (Kurstak et al.,

1978), has been shown to be identical to the rheumatoid arthritis nuclear antigen (Billins and Hoch, 1983). Hennesy and Kieff (1983) recognized two sets of nuclear antigens in EBV-transformed cells.

Carcinoembryonic antigen (Chapter IX, Section A,1) can be demonstrated after immunoblotting (Rittenhouse *et al.*, 1984) and can thus be used in cancer diagnosis.

4. Allergy

The usefulness of immunodotting for the rapid diagnosis of allergens which are involved in certain allergies has been shown elegantly by Walsh *et al.* (1984). Profiles of IgE reactivity among various allergens (Peltre, 1982; Weiss *et al.*, 1982) also proved very useful.

C. CONCLUSION

Immunoblotting is rapidly gaining an important place among immunoassays. It can, in principle, be used in three different ways: (i) after separation of immunoglobulins by electrophoresis or isoelectrofocusing; (ii) after immunization particular antibodies obtained can be analyzed on immunoblots; and (iii) in diagnosis antibodies specific to a certain polypeptide separated from others in a mixture by electrophoresis can be recognized.

7

The Use of Polyclonal or Monoclonal Antibodies in Enzyme Immunoassays

A. VARIATION IN IMMUNE RESPONSE

Polyclonal and monoclonal antibodies both have specific properties that make either of them ideal for a particular situation. The nature of these different properties should be well understood to use these antibodies optimally. Klinman and Press (1975) estimated that a BALB/c mouse contains 200 million B lymphocytes, of which 5 to 20% yield distinct clonotypes. Each plasma cell, derived from a B cell, produces antibody of one specificity, though the constant domains of the heavy chain may be interchanged by a recombination of heavy chain coding DNA with the same DNA sequences coding for the variable domains. Thus, the gene segments coding for the variable domain (V domain), which contains the antigen-binding site, can in the cellular development be first connected to the gene segments coding for the μ chain (thus producing IgM). After another recombination, the V domain gene may be linked to the gene segments coding for the constant domains of the γ chain (Leder, 1982; Tomagawa, 1983) resulting in the production of IgG. The sequences for the antigen-binding site are constructed in a hypervariable way prior to these recombinations so that a very large number of specificities are present among the various antibodies before the actual exposure of the animal to the antigen. Köhler (1976) and Klinman and Press (1975) estimated that a BALB/c mouse may contain 1000–7000 different B lymphocytes with the capacity of producing antibody to the same immunogen. Not surprisingly, some may have low and others high affinity depending on the degree of complementary between the epitope and the paratope, the number of secondary, attractive forces (noncovalency), and the steric factor (repulsive forces). From this pool of several thousand B lymphocytes, potentially giving rise to antibody-producing plasma cells, usually about 5–10 different clones respond fully. Briles and Davie (1980) suggested that this response is idiosyncratic and depends at most only partially on affinity-dependent maturation.

Genetic disposition is important for the immune response, a factor which is generally not taken into account sufficiently in immunodiagnosis. It can be demonstrated with inbred mice that one strain may respond strongly to an immunogen, whereas others react poorly. Harboe and Ingild (1983) demonstrated this important factor for the production of antisera in rabbits. The immune response (*Ir*) genes are situated in the major histocompatibility complex (Klein, 1979, 1982; Tonga *et al.*, 1983). The *Ir* gene products are designated as Ia antigens. In the mouse, 5 different *Ir* gene regions have been distinguished; the corresponding regions, with different compositions in man, are increasingly recognized (Gonwa *et al.*, 1983). On the other hand, for diagnostic purposes, it can be expected that large differences exist in the immune response of different individuals to the same immunogen, e.g., a virus.

B. THE CONCEPTS OF AFFINITY, AVIDITY, CROSS-REACTIVITY, AND SPECIFICITY

Affinity relates to the interaction of one monomeric paratope with one monomeric epitope. The affinity (K) can be defined in terms of concentrations

$$K = \frac{(\text{Ab}-\text{Ag})}{(\text{Ab})(\text{Ag})}(M^{-1})$$

which can be measured at equilibrium. This affinity constant is a measure of the standard Gibbs free energy of binding:

$$\Delta \text{G}^{\circ} = -RT \ln K$$

In practice, an antibody has two or more paratopes and antigens often have more than one epitope. It is then difficult to establish the exact value of K. Moreover, with polyclonal sera different types of heterogeneity may be encountered, i.e., antibodies with different intrinsic affinities or antibodies from different classes which may have a different "affinity bonus" since they have a different degree of polymerization ("valency"; e.g., IgG, IgA, IgM). A third level of heterogeneity is found among the antigens. These heterogeneities complicate the mathematical form of the algorithms used for data management. This heterogeneity causes curvature of Scatchard plots as recognized by Pauling *et al.* (1944). Most of the methods adapted for the determination of K assume a Gaussian error distribution (Karush, 1962) or a Sipsian distribution (Nisonoff *et al.*, 1975). Bowman and Aladjem (1963) used an integral transform theory to develop a method to calculate the distribution of binding energies. Siskind (1973) showed, however, that Gaussian distributions of energies cannot be accepted. A reasonable accuracy can be obtained by assuming a bimodal or trimodal distribution of a limited number of different binding energies (Klotz and Hunston, 1975).

The binding of an antigen to an antibody is generally short lived. It can be calculated from the dissociation constants (Kurstak *et al.*, 1984; Tijssen, 1985) that the half-life of an antibody–antigen is very often shorter than 1 sec, whereas the association constant is generally diffusion controlled and is quite similar for the various systems. Therefore, it is advantageous to envision the binding reaction as a constant association–dissociation process rather than long-term binding. This has very important implications for enzyme immunoassays.

In the first place, low to median affinity results in dissociation very rapidly after association, and thus more rapidly in the attainment of equilibrium. For example, during a washing step in an immunoassay, the buffer added does not contain any immunoreactant; low-affinity antibody will reach equilibrium rapidly with the antigen due to the rapid association–dissociation process, and thus dissociates to a degree determined by its affinity constant. Since the highest attainable antibody concentration adsorbed to a well of a microplate is about $10^{-7}\ M$, the affinity constant should be at least $10^{7}\ M^{-1}$, otherwise dissociation will increase considerably, and losses during washing will be considerable. From the equation above, it can be seen that

$$b = \frac{(\mathrm{Ab}-\mathrm{Ag})}{(\mathrm{Ab})_{\mathrm{total}}} = \frac{K(\mathrm{Ag})}{1-K(\mathrm{Ag})}$$

thus that the fraction, b, of antibody to which antigen is bound depends on K and the free (Ag). If K is smaller, then free (Ag) automatically will increase, i.e., if $K(\mathrm{Ag}) = 1$ then after one washing 50% of the antigen remains bound; however, if $K(\mathrm{Ag}) = 0.1$, 91% will desorb in the first washing.

In the second place, antibodies in an immune complex of several antigens and antibodies will not associate–dissociate in the same pattern. At the moment one antibody–antigen bond is disrupted, the complex is still held together by the other antibodies or antigens in the complex so that complete dissociation occurs at a much lower rate and reassociation is more rapid. This "affinity bonus" yields the avidity of the antibody and will be more pronounced for low-affinity antibodies (e.g., many of the IgM) than for high-affinity ones. This avidity is also important for lectins, the binding sites of which often have, separately, too low affinities.

Among the various antibodies present in a polyclonal antiserum only a low percentage (e.g., 5%) has high affinity. These antibodies are, however, dominant. Thus, in the production of monoclonal antibodies many of these antibodies will have low affinities which may yield disappointing results in immunoassays (Milstein, 1982).

Specificity can be defined as the difference in affinity of an antibody for the specific and nonspecific antigen. The affinity for the cross-reacting antigen is usually much lower. In a polyclonal antiserum, different clonotypes of antibodies contribute to the reaction with the antigen. However, each of these clonotypes generally has an affinity for different cross-reacting antigens, so that their combined cross-reactivity is lowered with respect to the specific response. This phenomenon is called "specificity bonus."

The cloning of monoclonal antibody-producing cells will also result in the loss (except on rare occasions) of the affinity and specificity bonus. The use of monoclonal antibodies demonstrated that it is necessary for immunoassays to distinguish the types of cross-reactivities that may be encountered. "Shared-reactivity" is used here to indicate cross-reactivity based on the presence of common epitopes in different antigens, whereas "true cross-reactivity" refers to the reactivity of the antibody to dissimilar epitopes. The affinity for the nonspecific epitope will generally be lower than that for the specific epitope. The concept of specificity is ambiguous in that it can be the opposite of shared-reactivity or of cross-reactivity. Specificity for the latter is defined as the ratio of affinities of the antibody of specific versus nonspecific antigens. If this ratio is lower than 10, then specificity is poor; if it is higher than 1000 then specificity is good.

C. THE CONSEQUENCES OF THE PRODUCTION OF MONOCLONAL ANTIBODIES FOR AFFINITY AND SPECIFICITY

The enhancement of affinity and specificity due to the presence of polyclonal antibodies is lost by the cloning of antibody-producing plasma cells. An obvious strategy to remedy this problem is to combine two or more different monoclonal antibodies. Since these will then recognize a different epitope, an increase in the apparent affinity (avidity) can be expected. This has been corroborated by the studies of Ehrlich *et al.* (1982). In contrast, Yolken *et al.* (1982) did not observe an affinity bonus after the mixing of different monoclonal antibody preparations.

Another most important consideration is the small fraction of antibodies which have high affinities. This may be less than 5–25%, and many monoclonal antibodies may, therefore, lack the affinity required (10^7 M^{-1}) for meaningful enzyme immunoassays. Two important factors are responsible for the maturation of antisera (increase in affinity without additional immunizations; e.g., Butler *et al.*, 1980). In the first place, immunoglobulin receptors on the lymphocytes (monomeric IgM and IgD) compete for the invading immunogen. This will be increasingly selective for lymphocytes that produce antibodies with higher affinities. Second, high-affinity antibodies produced will compete efficiently with the receptors on low-affinity antibody-producing lymphocytes (Eisen, 1980) and so prevent the latter from evolving into producers. However, in the case of monoclonal antibodies, the survival of hybrids is biased toward certain types of spleen cells. Claftin and Williams (1978) noted that it is quite possible that the best hybrids are not the best antibody-secreting plasma cells.

The shared-reactivity phenomenon is important for the specificity of monoclonal antibodies (Berzofsky and Schlechter, 1981). Antibodies do not recognize antigens but epitopes, and an epitope located on different antigens makes it impossible for a monoclonal antibody to distinguish these antigens. Polyclonal antisera, on the other hand, have the capacity of distinguishing such antigens since they recognize differently the two sets of epitopes on the two different antigens. Theoretically, however, this problem could be overcome by mixing the appropriate monoclonal antibodies. In fact, avoiding the monoclonal antibody recognizing the common epitope will avoid any nonspecificity. These considerations are of utmost importance for the diagnosis of infectious diseases (Wiktor, 1984). For example, many viruses undergo an "antigenic drift" which makes use of different panels of monoclonal antibodies imperative (Chapter IX). True cross-reactivity may also be decreased by the judicious selection of the monoclonal antibody with the highest ratio of affinities for the special and nonspecific epitope.

It is also possible that monoclonal antibodies are selected which are specific for a single strain of a microorganism and thus lack the ability to identify the different strains as required clinically (too specific; Yolken, 1982). Antibodies directed to common epitopes (on group antigens) should then be selected with the desired range of specificities (Gerhard *et al.*, 1978, Laver *et al.*, 1979; Zweig *et al.*, 1979).

A possible problem in the commonly used sandwich assay is that the antigen to be tested usually has only one epitope to which the monoclonal antibody can be bound. The solid-phase antibody binds this epitope and thus immobilizes the antigen. However, this antigen then lacks the capacity to bind the enzyme-conjugated antibody directed to the same epitope, and the tests give false negative results. Composite antigens, such as viruses, often have a large number of the same antigen (e.g., capsid protein), so this problem could then be avoided. In other cases this problem may be avoided by pooling or using different monoclonals (Yolken, 1982).

On the other hand, monoclonal antibodies offer the possibility of an experimental shortcut in that one monoclonal antibody is immobilized on the solid phase. The antigen and the labeled monoclonal antibody (specificity to another epitope) can then be added simultaneously since the two immune reactions take place on different sites of the antigen. Recently, Nomura *et al.* (1983) noted that sometimes a hook effect may be encountered with this otherwise appealing approach. An excess of labeled antibody may cover the antigen in such a way that the latter is prevented from reacting with the immobilized antibody. This phenomenon is dependent on the particular structure of the antigen and needs to be established for each particular antigen–antibody system.

A promising approach using the specificity of different monoclonal antibodies for neighboring epitopes on the antigen has been reported by Ngo and Lenhoff (1982). In this "proximal linkage enzyme immunoassay," no solid phase or washing steps are required. An enzyme labeled to one antibody produces a product that can be used as a substrate by the enzyme coupled to the second antibody recognizing a neighboring epitope. When these enzymes are homogeneously distributed in the solution the enzymatic reaction will be very slow (first order); however, bringing the enzymes close together, as achieved by the antigen (acting as a bridge), will increase the local concentration of the product used as a substrate by the second enzyme, and thus the antigen may be quantified. An example of a suitable enzyme pair is glucose oxidase as the first enzyme (producing hydrogen peroxide) and peroxidase as the second one (consuming hydrogen peroxide), and this can be quantified by appropriate oxidation of hydrogen donors. Since due to the proximity of glucose oxidase the local hydrogen peroxide concentration will be high, peroxidase can be measured easily.

An important problem which is encountered with polyclonal antisera is, as noted, the presence of anti-idiotype antibodies, anti-anti-idiotype antibodies, etc. On the one hand, these antibodies can be used for important therapeutic purposes (Koprowski *et al.*, 1984; Kennedy *et al.*, 1984), but, on the other hand, such antibodies may interfere in sandwich assays that use polyclonal antisera. Labeled anti-idiotype antibodies, which have specificities not necessarily only to the paratope, but also to other regions on the variable domain of the antibody, may react directly with the immobilized antibody and so yield false positive results. This phenomenon is, interestingly, observed more often with some systems (e.g., hepatitis virus) than with others. This problem may possibly be avoided by selecting antisera from different species to restrict anti-idiotypic activities.

D. UNUSUAL PROPERTIES OF MONOCLONAL ANTIBODIES

An antigen does not precipitate a monoclonal antibody directed to a single epitope, but polyclonal antibodies, which have antibodies to different epitopes, can do so readily. Some of the unexpected properties of monoclonal antibodies can be traced to this phenomenon (Milstein, 1982). An example is the lack of diffusion barriers (precipitation lines) in gels (Lachmann *et al.*, 1980). When polyclonal and labeled monoclonal antibodies are mixed and allowed to diffuse through the gel, they can saturate the epitope and pass through the precipitation line without dissolving it. Enzymatic revelation of enzyme-labeled antibodies will reveal their presence at both sides of the "diffusion barrier." It is also difficult to explain why pooling of three different monoclonal antibodies fails to simulate the properties of polyclonal antisera (Lachmann *et al.*, 1980). The underlying reason may be why Yolken *et al.* (1982) did not observe the expected affinity increase on pooling.

Stability of monoclonal antibodies is often unlike that of polyclonal antibodies. Polyclonal antisera are quite stable to a number of chemical or physical treatments. Even if one or two antibody species would be labile, this would hardly affect the detectability by this serum. If only 50% are stable, the loss of detectability would still be only one dilution step lower than originally (assuming similar affinities). Thus, this titer could be 1×10^6 in enzyme immunoassays. However, with monoclonal antibodies a 50% chance of a particular antibody species being stable could be disastrous, since the activity of each antibody (affinity) is lowered considerably.

Likewise, the optimal purification of monoclonal antibodies is unpredictable. Though many of these antibodies may be purified by methods used for polyclonal antibodies, very often they may not be. Also the purification of, e.g., a monoclonal IgG_1 molecule may be completely different from that of immunoglobulins of that subclass from polyclonal antisera. Some methods have proved useful for monoclonal antibodies, such as precipitation with caprylic acid (Steinback and Auban, 1979) and affinity chromatography on Blue-Sepharose (Böhm *et al.*, 1972) since they tend to lose less of certain antibodies. On the other hand, it is possible by careful examination of the purification procedure for each monoclonal antibody to find optimal conditions, and purities may then be achieved which are almost impossible to attain with polyclonal antisera.

It should be realized that monoclonal antibody is produced in malignant cells. The question of a potential biohazard should, therefore, be raised (Kaplan *et al.*, 1981), in particular with human myeloma cell lines. A careful investigation for reverse-transcriptase-bearing particles is warranted.

E. CONCLUSION

Monoclonal antibodies, in recent years, often produced quite disappointing results. Milstein (1982) pointed out that this does not justify statements that monoclonal antibodies are not as good as conventional antisera. However, the nature of monoclonal antibodies and of the immune reaction should be well understood. Appropriate mixing of monoclonal antibodies can often, but not always, mimic desired traits from polyclonal antisera. The problems noted with monoclonal antibodies are partially responsible for the relatively slow replacement of conventional polyclonal antisera by monoclonal antibody. It can be expected, however, that monoclonal antibodies will increase considerably the applicability of enzyme immunoassays. There is a very significant increase in the use of the immunoperoxidase technique due to the very specific reagents becoming commercially available. For example, in the past immunoperoxidase allowed spectrin to be detected in many cells. The availability of panels of monoclonal antibodies allowed Appleyard *et al.* (1984) to detect spectrin in different cells but with important differences in reactivities for the different monoclonal antibodies. These authors were, therefore, able to predict differences in composition and structure of these important proteins. On the other hand, different monoclonal antibodies may be used for purposes of double-immunoenzyme labeling (Mason *et al.*, 1982; Chapter 5) and the elimination of background staining. Monoclonal antibodies may also provide the tools to identify hitherto undiscovered constituents of cells.

The applications of monoclonal antibodies will be further discussed in the chapters dealing with the various fields of biology and medicine in which enzyme immunoassays are used.

Homogeneous Enzyme Immunoassays for the Detection of Drugs

A. INTRODUCTION

Homogeneous enzyme immunoassays rely on the modulation of the enzyme activity due to the interaction of the antigen (hapten) with the antibody (Fig. 4.6). The enzyme may be labeled to hapten, substrate, antibody, cofactor, or coenzyme. Alternatively, hapten may be labeled to molecules that modulate the enzyme reaction. In that case, antibody binding to the labeled hapten prevents this modulation; however, hapten present in the sample competes for the antibody and enhances modulation (Bacquet and Twumasi, 1984). Similarly allosteric effectors can be used as labels (e.g., methotrexase for dihydrofolate reductase; Boguslaski *et al.*, 1979). This group of assays is widely known under the trademark EMIT (Enzyme Multiplied Immuni Techniques; Syva Co., Palo Alto, C.A).

Three widely used enzymes for the detection of drugs are lysozyme (EC 3.2.1.17), glucose-6-phosphate dehydrogenase (G-6-P-DH; EC 1.1.1.49), and malate dehydrogenase (MDH; EC 1.1.1.37). The modulation may be achieved by several approaches. Sometimes, e.g., for lysozyme, the antibody blocks access of the large substrate (*M. lysodeikticus*) to the enzyme, whereas in other cases, e.g., for the dehydrogenases, conformational changes in the enzyme seem to cause modulation. For other enzymes, e.g., β-galactosidase, substrate may be made macromolecular by linkage to a high-molecular-weight polymer.

Haptenation of MDH with thyroxine results in a 10-fold lower enzyme activity. However, antibody reacting with thyroxine reactivates the enzyme (7- to 9-fold). Ullman *et al.* (1979) observed that covalent binding of two thyroxines per enzyme unit is necessary to produce this effect. Antibodies reacting with triiodothyronine, which strongly resembles thyroxine, on MDH inhibit the enzyme (Kabakoff and Greenwood, 1982). It seems that conjugates contain both inhibitable and activatable fractions.

B. ASSAY DESIGN AND PROTOCOL

A prerequisite for homogeneous enzyme immunoassay is the availability of antibodies to the drugs or drugs of abuse. The carrier proteins most often chosen are bovine serum albumin, bovine IgG, or keyhole limpet hemocyanin (Albert *et al.*, 1978).

The conjugation of the hapten to the enzyme follows the methods discussed in Chapter III, Section B. Frequently, *N*-hydroxysuccinimide or mixed anhydrides are chosen. Optimal conjugation is probably more pronounced for the homogeneous than for the heterogeneous assays and should be determined empirically. Two factors are important in this respect: degree of inhibition due to labeling (L) and the degree of modulation possible with antibodies (M). A high M/L ratio will result in high detectabilities (other assay influences being constant).

In the next step, the optimum antibody concentration is determined for the desired concentration range for the detection of the drug. A given concentration of enzyme–conjugate is incubated with different amounts of antibody to determine the maximum modulation possible. Next, several combinations of conjugates–antibody are tested over the therapeutic range. The combination of high modulation with high sensitivity (i.e., dR/dC; R = response, C = concentration of drug) over the therapeutic range is established. Assay specificity (lack of cross-reactivity or shared reactivity) in EMIT assays can be established using a fixed concentration of the specific drug to be detected and by adding variable amounts of cross-reactants. The cross-reactant can selectively block the less specific antibodies and desensitize the assay for these cross-reactants (Kabakoff and Greenwood, 1982). This method is routinely used in the production of commercial assays, e.g., for procainamide (*N*-actylprocainamide is the cross-reactant) *N*-acetylprocainamide, theophylline, and tidocaine (Kabakoff and Greenwood, 1982).

The ‘‘stat’’ procedures are widely used for assays in which detectability is not the prime factor, but speed and simplicity are desired (urine drug screening tests, serum toxicology assays, therapeutic drugs except digoxin). For the urine assays, lysozyme is used with *M. lysodeikticus* as substrate (at pH 6.0) with 50 μl sample at 37°C. The decrease in light scattering ("absorbance") at 436 nm is measured over 40 sec. For serum assays, G-6-P-DH is used with glucose 6-phosphate and NAD as substrate (pH 8.1) and 50 μl 1:6 diluted serum. The absorbance change at 340 nm is measured for 30 sec (assay temperature is 30°C). Calibration curves change very little during a working day and single measurements can thus be made.

The batch methods (Chang *et al.*, 1975) have higher detectabilities but require longer incubation periods and larger sample sizes. A calibration curve should be run with each batch of samples. The samples are pretreated, e.g., to liberate T_4 from thyroid-binding globulin or to destroy endogenous enzyme activity. Syva Co. includes Liplex R in the tyroxine assay to sequester free fatty acids. Sodium oxamate is included for the automated digoxin in assay to reduce lactate dehydrogenase interference. The pretreatment time takes 5 or 10 min, the antibody incubation 5 or 15 min, and the addition of enzyme substrate and kinetic incubation 15–30 min (initial and final absorbance measured). A disadvantage of this method is its laboriousness.

A third type assay, the end-point assay, avoids some of these problems while remaining sensitive. The EMIT digoxin assay is performed at room temperature. After pretreatment with pepsin, antibody is added, incubated for 10 min, and enzyme reagent is added. The reaction is stopped after 90–190 min by adding antibody or sodium borate to the enzyme (G-6-P-DH). The product formation can then be measured (340 nm).

Automation is convenient for these assays (e.g., with the ABA-100, Centrifichem, Gemsaec, Rotochem, Gilford 3500, LKB2074, ADA, AA11, KDA, SMA, SMAC and Autochemist; Kabakoff and Greenwood, 1982; Greenwood and Chandler, 1980).

C. QUALITATIVE AND SEMIQUANTITIVE DETECTION OF DRUGS

1. Detection of Drugs of Abuse

Some commercial EMIT assays have a wide spectrum (e.g., the barbiturate assay; Walberg, 1974) whereas others (e.g., benzoyl ecgonine; Mulé *et al.*, 1977) are highly specific.

Urine assays for drugs of abuse (using lysozyme) have been reviewed by Curtis and Patel (1978). However, lysozyme may sometimes be present in urine, so blanks should be run to exclude false positives. Moreover, the pH or salt concentration of the urine and the presence of lysozyme inhibitors may affect the enzyme activity. These limitations may be overcome to some extent by using an indirect method (Slighton, 1978) or using another enzyme (Tom *et al.*, 1979). EMIT assays for drugs of abuse are listed in Table 8.1 (adapted from Kabakoff and Greenwood, 1982).

The benzodiazepine assay uses enzyme-labeled oxazepam (a metabolite of all benzodiazepines) and is highly specific. Benzodiazepine is one of the most abused drugs (Curtis and Patel, 1978).

The amphetamine assays have high cross-reactivity resulting in frequent false positives. Cross-reacting drugs may be found among those prescribed for cold remedies, direct medication, or for other reasons (Curtis and Patel, 1978).

2. Detection of Therapeutic Drugs

a. Bronchodilating Drugs

Theophylline is a primary drug of choice for the management of acute and chronic asthma (Weinberger and Hendeles, 1979).

Antisera are raised against theophylline derivatized at the 3-*O*-methyl position. A number of related xanthines (e.g., caffeine) may be present in

TABLE 8.1
Homogenous Enzyme Immunoassays for Drugs of Abuse

Assay	Calibration drug	Detectability (μg/ml)
Opiate	Morphine	0.5
Amphetamine	Amphetamine	2.0
Cocaine metabolite	Benzoyl ecgonine	1.6
Barbiturate[a]	Secobarbital	2.0
Propoxyphene	Propoxyphene	2.0
Benzodiazepine[a]	Oxazepam	0.7
Methadone	Methadone	0.5
Cannabinoid	11-nor-A^8-THC-9-carboxylic acid[b]	0.05[c]
Phencyclidine[a]	Phencyclidine	0.15[c]

[a] Based on G-6-P-DH.
[b] THC, tetrahydrocannabinol.
[c] Based on MDH.

the host, however, interference is generally very low (Koup and Brodsky, 1978; Oellerich *et al.*, 1979; Weidner *et al.*, 1979). Dietzler *et al.* (1980, 1983) applied the Hill equation for the determination of the theophylline concentration. This mathematical approach is quite similar to the logit procedure. Obtaining the exact value of maximum response to prevent pronounced curvature of the standard line is of utmost importance.

b. Cardioactive Drugs

EMIT assays have been developed for several cardioactive drugs (digoxin, lidocaine, procainamide, *N*-acetylprocainamide, propanolol, quinidine, and disapyramide; Walberg and Won, 1979; Cobb *et al.*, 1977; Scoggin *et al.*, 1978; Kabakoff and Greenwood, 1982).

Two separate assays should be performed to distinguish between procainamide and *N*-acetylprocainamide, while with HPLC they can be measured in a single run. No such problems are encountered with lidocaine or disapyramide.

Pretreatment with pepsin is necessary for endogenous enzyme denaturation for the digoxin test. Moreover, this pretreatment has the advantage that endogenous proteins which interfere in radioimmunoassays of this drug (Painter and Vader, 1979) are eliminated.

3. Antiepileptic Drugs

EMIT assays have been developed for serum monitoring of the major anticonvulsants (phenytoin, phenobarbital, premidone, carbamazepine, ethozusimide, and valproic acid). In particular, valproic acid is a very small hapten (di-*n*-propylacetic acid), but antibodies raised against it have high specificity (Schottelius, 1978; Izutsu *et al.*, 1979; Kabakoff and Greenwood, 1982).

Saliva drug measurements are also feasible for the estimation of free drug concentrations (Danhof and Breimer, 1978).

4. Methotrexate

Reliable establishment of methotrexate concentrations is not of primary importance in achieving minimum therapeutic levels, but is necessary to prevent or to preduct toxicity when administering high doses (i.e., to monitor the clearance of the drug from the body; Evans *et al.*, 1979). In general, the sample has to be diluted considerably to bring it in the range of the assay's calibration curve.

Specificity has been investigated, particularly with respect to the metabolites 1,4-diamino-N^{10}-methylpteroic acid and 7-hydroxymethotrexate (Donehower *et al.*, 1979; Paxton, 1979; de Porceri-Morton *et al.*, 1980). Cross-reactivity with the pteroic acid compound may be considerable, leading to escalation of leucovorin rescue (Stoller *et al.*, 1977). Leucovorin or other folates do not cross-react.

5. Antibiotics

Enzymatic assays for gentamycin and tobramycin have been developed and evaluated for clinical applications (Holmes and Sandford, 1974; Phaneuf *et al.*, 1980; Sanders *et al.*, 1979). Tobramycin is reportedly more effective than gentamycin against *Pseudomonas aeruginosa* infections. Monitoring is indicated by ototoxicity and nephrotoxicity (peak levels 10 through 2 μg/ml). The EMIT assay (Syva Corp., Palo Alto, Ca) requires less than 10 μl of serum and is completed in less than 1 min. The conjugate used is tobramycin–G-6-P-DH enzyme. The gentamycin assay is modeled closely after the tobramycin assay (Kabakoff *et al.*, 1978). Antibodies are obtained by immunizing sheep with a gentamicin–bovine serum albumin (BSA) conjugate.

In their method Phaneuf and colleagues (1980) used a bacterial glucose-6-phosphate dehydrogenase enzyme to which gentamycin had been conjugated. The active site of the enzyme was adjacent to the bound aminoglycoside. Anti-gentamycin antibody links to the gentamycin adjacent to the enzyme. The assay follows the conversion of nicotinamide adenine dinucleotide to reduce micotinamide adenine dinocleotide which occurs when the enzyme is able to act on a substrate. Free gentamycin from a sample compete for antibodies with drug bound to the enzyme. When less antibody blocks the active site of the glucose-6-phosphate dehydrogenase, the activity of the enzyme is increased. Of a sample 50 μl was diluted with 250 μl of 0.055 *M* Tris buffer, pH 8.0. Subsequently, 50 μl of the first dilution was diluted with 250 μl of buffer to which was added 50 μl of substrate and antibody and 250 μl of buffer. Finally, 50 μl of gentamycin–enzyme conjugate plus 250 μl of Tris buffer were added. Absorbance of reaction was measured at 340 nm.

With the enzyme immunoassay for gentamycin only 30 min is necessary to return an assay result to the physician (Phaneuf *et al.*, 1980).

Recently, Fujiwara *et al.* (1984) developed a heterologous enzyme immunoassay to quantify puromycin aminonucleoside (PA). Their method is based on the use of anti-puromycin antibody and used β-D-galactosidase-labeled PA conjugate prepared via N-(m-maleimidobenzoyloxy) succinimide. This EIA was found to be 20 times more sensitive than the homologous EIA for PA with anti-PA antibody and PA-β-D-galactosidase conjugate. The heterologous enzyme immunoassay was free from purine or pyrimidine analogs interference and drug levels were determined in rat tissue at a dose of 15mg/kg.

6. Hormones and Vitamins

The enzyme-linked immunosorbent assays are now available for the measurement of several hormones. Recently a micro-ELISA test for detection of insulin antibodies in patients serum was developed by Wilkin *et al.* (1985). Insulin-binding antibodies in high concentration could induce insulin resistance, allergy, and local lipoatrophy (Reeves *et al.*, 1980). To measure these antibodies in the serum of diabetics a solid-phase micro-enzyme-linked immunosorbent assay of Wilkin *et al.* (1985) is an alternative to radioassay. ELISA provides a specific and reproducible method of detecting insulin antibodies in sera, and is not subject to the restrictions associated with the radioisotopes. The specificity of ELISA was demonstrated by substituting purified human gamma globulin for the test serum and glucagon for the insulin. The influence on ELISA of endogenous insulin in the test serum was examined by measuring antibody binding before and after extraction of insulin.

The gonadotropin hormone is synthetized and secreted in females as early as 170 hr after fertilization and is used as an index of pregnancy. Enzyme immunoassays have today replaced the previous methods based on erythrocyte-agglutination or latex-agglutination. Recently Gupta *et al.* (1985), developed a two-sided sandwich enzyme immunoassay for human chorionic gonadotropin (hCG) employing monoclonal antibodies directed against β- and α-subunits. Monoclonal anti-β-hCG antibody was used for coating microtitration plates and monoclonal anti-α-hCG antibody labeled with horseradish peroxidase was used as tracer. Their assay could detect up to 1 μg hCG/ml. A higher sensitivity enabling detection up to 0.25 μg hCG/ml was possible in the sandwich enzyme immunoassay with the use of biotin-avidin interface.

A rapid and sensitive enzyme-linked immunosorbent assay was also developed by Maeda *et al.* (1985) for mass-screening of neonatal congenital hypothyroidism using dried blood samples on filter paper. The assay is carried out on microtiter plates without extraction or centrifugation steps. The detection limit of the proposed assay is 5 pg/disc/well, equivalent to 1.25 μg/liter of whole blood or 2.5 μg/liter of serum.

Rapid analysis of vitamins by immunoassays have been reported for water-soluble vitamins (Endres *et al.*, 1978; Howe *et al.*, 1979). In their recent work, Brandon *et al.* (1985) developed homogenous immunoassays for individual vitamins of B_6. The detection limits of 0.1 μg for the enzyme immunoassay compare favorably with described methods employing high performance liquid chromatography, which have a detection limit of about 1 μg (Gregory *et al.*, 1981). ELISA should provide useful tests for the quantitation of vitamins, for the analysis of foodstuffs, and for research in the areas of chemical nutrition.

The Value of Enzyme Immunohistochemistry in Pathology, Oncology, and Histology

Enzyme immunohistochemistry was foreshadowed by the greatly successful immunofluorescence methods (Coons *et al.*, 1950). Immunofluorescence procedures were developed for the examination of biopsy material for deposits of immunoglobulin and complement resulting from immunopathological events. This approach proved particularly useful for the diagnosis of renal and skin diseases (Mellors *et al.*, 1957). Subsequently, the indirect immunofluorescence technique enabled the detection of anti-tissue antibodies in patients' sera (e.g., antinuclear factors in systemic lupus erythematosus sera; Holborow *et al.*, 1957).

The initial advantages of enzymes over fluorescent tracers were different from those now known. Initially, the trouble and expense of dark-field ultraviolet illumination, the obscuring effect by autofluorescence of the tissue, and the fading of fluorescein staining were the major disadvantages of fluorescent over enzyme tracers. These are no longer problems due to advances in fluorochrome-label methodology (illumination at specific visible-light wavelengths can be used, addition of *p*-phenylenediamine to glycol mounting fluid retards fading). However, enzymes are themselves antigenic, and this property can be exploited to enhance both specificity and sensitivity (Heyderman, 1979), and ultrastructure investigations are possible.

A. ENZYME IMMUNOASSAYS IN DIAGNOSTIC PATHOLOGY

Immunocytochemical techniques have the advantage of specificity for certain products over conventional stains. The localization of products indicative of a pathological state has, therefore, assumed a role of growing importance in diagnostic histopathology. The majority of the applications in this area are found in oncology, since tumors produce a wide array of products such as tumor antigens, hormones, oncofetal antigens, milk proteins, and immunoglobulins (DeLellis *et al.*, 1979).

Immunocytochemical studies also proved valuable for the diagnosis of renal disease (Zollinger and Mihatsch, 1978) and some dermatological disorders (Cooperative Study, 1975).

Immunoenzymatic applications for the identification of immunoglobulin and complement components for the diagnosis of immunologically mediated renal diseases offer an advantageous alternative to the widely used immunofluorescence methods (Tubbs *et al.*, 1980). Tubbs *et al.* (1980) employed the immunoperoxidase technique and periodic acid–Schiff counterstain on fresh-frozen tissue sections. Permanent preparations are obtained and the distribution of glomerular, vascular, interstitial, and tubular immune complexes can be correlated directly to histologic features observed on the same slide. A distinction could be frequently made between epimembranous, intramembranous, and subendothelial deposits. Sinclair *et al.* (1981) used paraffin sections of routinely processed renal biopsies (fixed in neutral buffered formol saline for 6–18 hr) and compared the indirect and the PAP immunoperoxidase methods. Though the indirect method was somewhat less sensitive, a similar staining could be obtained using higher concentrations of conjugate, thus avoiding the more complex methology of the PAP method. Predigestion by protease VII (Sigma P5255) at 0.05% in PBS (pH 7.3) for 20 min at 37°C effectively revealed the antigens without damaging the tissue.

α_1-Antitrypsin levels in the plasma often increase in cases of liver disease (Carlson and Eriksson, 1980). Carlson *et al.* (1981) applied the immunoperoxidase method on needle liver biopsies using rabbit antiserum monospecific for α_1-antitrypsin. Diffuse or granular intracellular α_1-antitrypsin deposits were observed in a minority of biopsies. The method enabled the correct identification of intrahepatocellular α_1-antitrypsin globules in the majority of patients with phenotype MZ and Z and was found to be more sensitive than PAS staining.

Recently, the detection and distribution of isoenzymes of lactate dehydrogenase (LDH) in patients with acute myocardial ischeminic syndromes have provoked interest. Herscher *et al.* (1984) demonstrated by the immunoperoxidase method that LDH-1 can be demonstrated in the normal myocardium and that loss of LDH-1 from necrotic myocardium can be demonstrated as early as 3 hr after coronary occlusion. LDH localization in tumor cells may be responsible for the elevated LDH in the serum of patients with some malignancies (Murakami and Said, 1984).

B. ENZYME IMMUNOHISTOCHEMICAL DETECTION OF ANTIGENS IN ONCOLOGY

Immunoperoxidase allows the simultaneous evaluation of morphology and the detection of specific antigens. Tissues which are routinely fixed and embedded can often be used for current as well as retrospective (e.g., 25 years after embedding) evaluation. The staining is permanent with some substrates and semipermanent with others, allowing the investigator to study functional aspects of antigens or abnormal cell types. In Table 9.1 applications pertaining to oncology and reported findings are listed.

Immunoperoxidase assays and monoclonal antibodies are now currently used for the characterization and diagnosis of tumor-associated antigens (Friedman *et al.*, 1985) and to define the epitopes on carcinoembryonic antigens in human colon carcinomas (Muraro *et al.*, 1985). Avidin–biotin–peroxidase method is also applied to the study of antigens expressed on gastric, ovarian, colon, endometrial, and cervical human carcinomas (Prat *et al.*, 1985).

1. Carcinoembryonic Antigen (CEA)

CEA is localized in a high percentage of adenocarcinomas of the large intestine. CEA can be detected relatively often in patients with other diseases or even in healthy individuals (Goldenberg *et al.*, 1978; Stenger *et al.*, 1979), though Gold *et al.* (1968) and Denk *et al.* (1973) have reported the absence of this antigen in normal tissue. Nevertheless, the detection of CEA is clinically useful (Burtin *et al.*, 1978).

Immunoperoxidase is a very useful technique for retrospective studies of CEA. Harrowe and Taylor (1981) could detect this antigen in specimens embedded 25 years previously.

TABLE 9.1
Detection of Antigens by Enzyme Immunohistochemistry in Oncology

Immunoglobulins
- Detection of extracellular deposits of immunoglobulins in renal biopsies
- Detection of malignant lymphomas of B origin
- Differentiation between multiple myeloma and macroglobulinemia
- Identification of undifferentiated neoplasms as malignant lymphoma
- Distinction between reactive and neoplastic plasma cell proliferations

Myoglobin
- Characterization of rhabdomyosarcomas and rhabdomyoblastic differentiation (e.g., teratocarcinoma)

Lysozyme
- Identification of cells of myeloid, monocytic, or histiocytic derivation in leukemias, lymphomas, or other infiltrates
- Identification of myeloblastomas

Keratin proteins
- Identification of neoplasms with squamous differentiation
- Transitional cell carcinomas and mesotheliomas
- Identification of spindle cell variant of carcinomas and mesotheliomas
- Identification of their absence in sarcomas and lymphomas

α-Fetoprotein
- Present in hepatocellular carcinoma
- Present in ovarian and testicular germ cell tumors (particularly embryonal and endodermal sinus types)

Glial fibrillary acidic protein
- Identification of glial cells in reactive and neoplastic processes
- Identification of glial nature of anaplastic intracerebral neoplasms

Carcinoembryonic antigen
- Identification of absence in neoplastic cells of sarcomas and lymphomas
- Identification of presence in adenocarcinoma of colon and other organs
- Provision of evidence of epithelial differentiation in undifferentiated neoplasma

Estradiol
- Present in granuloma and theta cell tumors

Testosterone
- Present in Sertoli–Leydig cell tumors

Calcitonin
- Identifies medullary carcinoma of thyroid and C cell hyperplasia of thyroid

Chorionic gonadotropin
- Identification syncytiotrophoblasts and choriocarcinomatous elements of ovarian and testicular germ cell tumors

Terminal deoxynucleotidyltransferase
- Classification of non-T, non-B, and T cell lymphoblastic leukemias

Hormones
- Identification by hormones produced by islet cell, carcinoid, and pituitary tumors
- Identification of ectopic hormone production in neoplastic cells

Tissue manipulation greatly influences CEA detectability. Goldenberg *et al.* (1978) detected CEA in frozen ethanol-fixed specimens but not in formalin–paraffin-processed specimens. However, the PAP method with its greater detectability enabled Wagener *et al.* (1978) to obtain CEA-positive staining in conventionally processed specimens of the normal mucosa. Primus *et al.* (1982) confirmed that the detection and localization of CEA by the immunoperoxidase procedure are more successful with the PAP method than with the covalently labeled antibodies, which in turn proved more sensitive than the triple-bridge procedure (Sharkey *et al.*, 1980). However, Primus *et al.* (1982) noted that ethanol fixation prior to paraffin embedding increases the specific staining intensity over those in which formalin fixation was used. This phenomenon has also been observed for other antigens. For example, the ethanol fixation method of Sainte-Marie (1962) enhanced the staining of α-fetoprotein (Kuhlmann, 1979). Protease treatment of paraffin-fixed specimens did not enhance the CEA staining intensity, suggesting that low staining is due to incomplete fixation rather than masking of the antigens.

Frozen sections yield a higher background with immunoperoxidase than with immunofluorescence (Wagener *et al.*, 1978), particularly at low serum dilutions, demonstrating that proper dilutions of anti-CEA antisera are very important. In contrast Lindgren *et al.* (1982) observed that background staining is minimal when monoclonal antibodies to CEA were used (Lindgren *et al.*, 1982). In Wagener's method, surgical and autopsy specimens were fixed in Bouin's fixative or in 10% buffered formalin (pH 7.4) and embedded in paraffin by the usual techniques. Incubations for 15 min for the immunohistochemical steps were at 37°C, and the sera were diluted in 50 m*M* Tris–HCl saline, pH 7.6. The dilutions used for optimum results were anti-CEA antiserum (from rabbit), 1:200; anti-rabbit γ-globulin, 1:40; and the PAP (commercial preparation), 1:50.

The dominant feature of the characteristic staining pattern of CEA (at the light microscopic level) is a linear labeling of the apical poles of cells lining the glandular lumen and mucosal surface. The weaker cytoplasmic staining is apparent only in the upper portion of the cell. Huitric *et al.* (1976) localized CEA within the mucigen droplet of normal goblet cells in the colon. Isaacson and Judd (1977) reported finding CEA in the small intestine (light and electron microscopic levels), but this was not substantiated by subsequent studies by Primus *et al.* (1982).

Mesothelial tumors have been shown to contain keratin proteins (Schlegel *et al.*, 1980) but to lack CEA (Wang *et al.*, 1979). Based on these observations, Said *et al.* (1982) concluded that adenomatoid tumors (epididymus, testis, uterus, fallopian tube) are of mesothelial origin due to the strong keratin staining and the lack of CEA or factor VIII staining. The potential value of CEA detection is often challenged since CEA is often demonstrable in normal colon cells. Using the immunoperoxidase technique O'Brien *et al.* (1981) investigated particular staining profiles and their relationships to benign, premalignant, and malignant colorectal tissues. Cross-reactivity of anti-CEA antibodies to related proteins is a potential source of positive staining and antisera should be absorbed with the normal antigens. Nevertheless, this could be a problem with very sensitive methods.

O'Brien *et al.* (1981) observed that (i) localization of CEA to glycocalyx or surface epithelial cells is a normal finding (normal colon or in mucosa adjacent to infiltrating carcinoma); (ii) strongly positive surface and intraluminal staining is not a reliable diagnostic criterion; and (iii) absence of CEA in gland-forming carcinoma makes diagnosis of colorectal carcinoma unlikely. The strong staining in colorectal adenocarcinoma contrasts with adenocarcinomas of the prostate, endometrium, ovary, pancreas, and biliary. In a preliminary report, Speers *et al.* (1983) compared immunohistochemically detectable levels of CEA in adenocarcinomas and microglandular hyperplasias of the endocervix. Their results suggest that the presence of CEA is indicative of malignancy.

The immunoblotting method using peroxidase-conjugated immunoglobulins was recently applied to define the epitopes of carcinoembryonic antigen. The CEA-specific monoclonal antibodies used reduce the incidence of false positive results in clinical diagnosis (Haggarty *et al.*, 1986)

2. Prostate-Specific Antigen and Prostatic Acid Phosphatase

Primary prostatic carcinomas can be demonstrated by immunoperoxidase staining for prostatic acid phosphatase (Table 9.2). Virtually all carcinomas contain this antigen, but it is absent from normal cases (Nadji, 1980). A disadvantage is that staining for prostatic acid phosphatase does not correlate with the differentiation of the tumor (Nadji *et al.*, 1980).

M. C. Wang *et al.* (1979) described a prostate-specific antigen, distinctive from acid phosphatase. This antigen can, however, be detected in benign prostatic hyperplasia, whereas absence of staining is not indicative of the absence of prostatic carcinoma (Stein *et al.*, 1982).

TABLE 9.2
PAP Method for Localization of Prostatic Acid Phosphatase[a]

1. Deparaffinize and hydrate tissue sections
2. Rinse for 2 min with PBS
3. Place in 2.28% periodic acid (freshly made) for 5 min[b]
4. Wash with running water for 5 min
5. Place in 0.02% freshly prepared sodium borohydride for 2 min[b]
6. Wash with running water for 5 min
7. Incubate for 15 min at 37°C with normal goat serum (1:10)
8. Wash and incubate with rabbit antiserum to prostatic acid phosphatase (Eureka Laboratories, Sacramento, Ca) at 37°C for 15 min [dilution (1:512)]
9. Wash with PBS (3 × 3 min)
10. Incubate with goat antisera to rabbit IgG (1:16) at 37°C for 15 min
11. Wash with PBS (3 × 3 min)
12. Incubate with PAP (Cappel Laboratories, Cochranville, Penn) for 15 min at 37°C (dilution 1:32)
13. Reveal enzyme (section)
14. Wash for 3 min in running water
15. Counterstain with Harris hematoxylin for 2 min
16. Dehydrate, clean, and mount

[a] According to Ansari *et al.* (1981).
[b] Steps introduced to inhibit background staining (Heyderman and Neville, 1977).

Ansari *et al.* (1981) noted an unusual case of carcinoma of prostate with carcinoid-like areas in the prostate as well as in the metastasized areas. Fontana-Masson stain and electron microscopy (to test the presence of neurosecretory granules) failed to confirm the possibility of primary carcinoid of the prostate. Immunoperoxidase staining revealed that this tissue was not true carcinoid but rather prostatic carcinoma with a carcinoid-like pattern, demonstrating the great usefulness of this technique. However, unequivocal assessment of the specificity of the antisera is of primordial importance.

Recently, Yamaura *et al.* (1985) produced monoclonal antibodies against human prostatic acid phosphatase and determined them by ELISA assay and immunohistological staining. Monoclonal antibodies are now used in the immunoperoxidase assay to characterize human prostate tissue antigens (Starling *et al.*, 1986). These authors isolated two monoclonal antibodies which bind to new prostate organ- and tumor-associated antigens. These developments are very useful for characterization of antigens and for diagnostic purposes.

3. Detection of Terminal Deoxynucleotidyltransferase in Acute Leukemia

Among the modern methods used to classify acute leukemia, terminal deoxynucleotidyltransferase (EC 2.7.7.31; TdT) has become an important marker for primitive lymphoblasts. It is positive in most cases of non-T, non-B, and T cell lymphoblastic leukemias (Coleman *et al.*, 1976). This DNA polymerase adds deoxynucleotidyl bases to the 3-OH terminus of primers without a template (Beutler and Blume, 1979). Bollum (1975) raised antibodies in rabbit against bovine TdT which cross-reacts with human TdT. These antibodies are commercially available from BRL Labs or P-L Biochemicals.

The biochemical assay is time consuming and requires a large number (10^7) of fresh cells. Immunofluorescence was, therefore, used in routine methanol-fixed smears or frozen sections, but this method has the general disadvantages mentioned earlier (e.g., requires an epifluorescence microscope). Several groups have adopted the immunoperoxidase test to replace these biochemical and immunofluorescence methods. Ho *et al.* (1982) optimized the PAP technique for this purpose and compared it to biochemical and immunofluorescence methods to detect TdT. Optimal results were obtained when primary antibodies (1:4 dilution) were incubated with the smears for 24 hr at 4°C, but 60 min at 37°C (1:2 dilution of serum) is equally practical. This method correlates excellently with the biochemical assays. Background staining caused by endogenous peroxidase is blocked by a prior incubation of the preparation with 0.3% H_2O_2 for 15 min (Stass *et al.*, 1982), although, according to Hecht *et al.* (1981), this enzyme does not interfere in this test. Stass *et al.* (1982) observed with the indirect immunoperoxidase test the same detectability as with the immunofluorescence test, whereas Ho *et al.* (1982) and Hecht *et al.* (1981) observed a higher detectability using the PAP method.

Alternatively, Halverson *et al.* (1981) used B-5 fixative (221 m*M* mercuric chloride, 152 m*M* sodium acetate, and 4% formaldehyde) for 30 min at room temperature. The peripheral blood sample is defibrinated by stirring with two glass Pasteur pipets, and, after gelatin sedimentation (i.e., after adding 3% w/v gelatin to 3 parts of whole blood, mixing and incubation for 1 hr at 37°C), washed, and resuspended to 10^6 cells/ml; the cells were either centrifuged (200 μl of suspension per slide) prior to fixation or fixed directly and embedded in paraffin.

Treatment of sections with DNase I for 4–6 hr (Janossy *et al.*, 1980) or with double-bridge methods (Halverson *et al.*, 1981) enhanced the perox-

TABLE 9.3
TDT–PAP Method[a]

1. Melt paraffin sections overnight at 60°C and rehydrate by washing three times with xylol and a graded series of alcohol
2. Preheat with DNase I for 5 hr at room temperature
3. Block endogenous peroxidase
4. Incubate for 10 min with normal swine serum (1:20)
5. Wash
6. Incubate with primary anti-TdT antiserum overnight
7. Wash
8. Incubate with bridge-antibody (anti-IgG at 1:100) for 30 min
9. Wash
10. Add PAP at 1:200 for 30 min
11. For bridge method, repeat steps 8–10
12. Wash and stain with DAB for 3–5 min
13. Treat for 5 min in Lugal's (1% iodine and 2% potassium iodide) rinse in tap water, dip for 60 min in 5% sodium hyposulfate, and rinse in tap water before counterstain is added; use 0.1% Matheson light green in 0.1% acetic acid or Mayer's hematoxylin as counterstain
14. Mount

[a] According to Halverson *et al.* (1981).

idase staining. The outline of the method used by Halverson and his collaborators is presented in Table 9.3.

Halverson *et al.* (1981) studied TdT in rat, calf, and human thymus and in neoplastic tissue, using routine smears or paraffin-embedded tissues. They confirmed that antibody to calf thymus TdT cross-reacts with rodent and human TdT (Bollum, 1975; Kung *et al.*, 1978). TdT-positive cells are primarily located in the cortical region of the normal thymus, but with certain lymphoproliferative diseases of T cell origin, TdT cells are detected in peripheral blood. However, paraffin-embedded tissue required DNase treatment and incubation overnight of the primary antibody.

4. Epithelial Membrane Antigen (EMA)

The globule membrane of milk lipids, secreted by lactating mammary epithelial cells, is acquired from the luminal surface of the cell when the globule passes from the cytoplasm to the lumen. These globule membranes contain a component, EMA, present in surface membranes of many epithelia, except squamous epithelia and the proximal convoluted tubule of the kidney (Sloane and Ormerod, 1981). Squamous carcinomas, however, are positive, whereas many other malignant tumors tend to have increased quantities of this antigen (Fig. 9.1). Cytoplasmic staining is particularly evident in cases in which tumor cells are isolated [e.g., nasopharyngeal carcinoma (Fig. 9.2), metastatic breast carcinoma (Fig. 9.3), primary breast carcinoma (Fig. 9.4)]. Adjacent cell membrane staining is often observed in positive malignant tumors as is staining of intracytoplasmic lumina in some malignant tumors (particularly breast carcinoma). Staining becomes more unpredictable with increasing degrees of anaplasia.

Though EMA is often present in normal epithelial and mesothelial tissues, the staining pattern in the tumors arising from them is different. The antigen can be demonstrated in conventional formalin-fixed paraffin-embedded sections. The technique may be particularly valuable in distinguishing anaplastic carcinoma from malignant lymphoma. The identification of minute metastasis in organs such as the liver and the bone marrow is also facilitated.

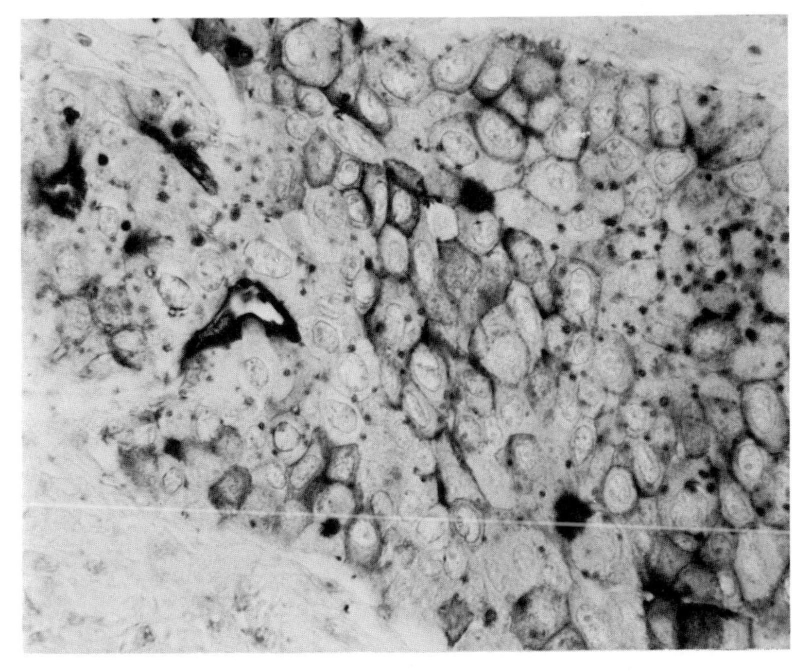

Fig. 9.1. Strong luminal membrane (left) and adjacent cell membrane (center and right) staining in infiltrative ductal carcinoma of the breast. Throughout, numerous small lumina in the cytoplasm are stained positively. Courtesy of Dr. J. P. Sloane with permission to reproduce.

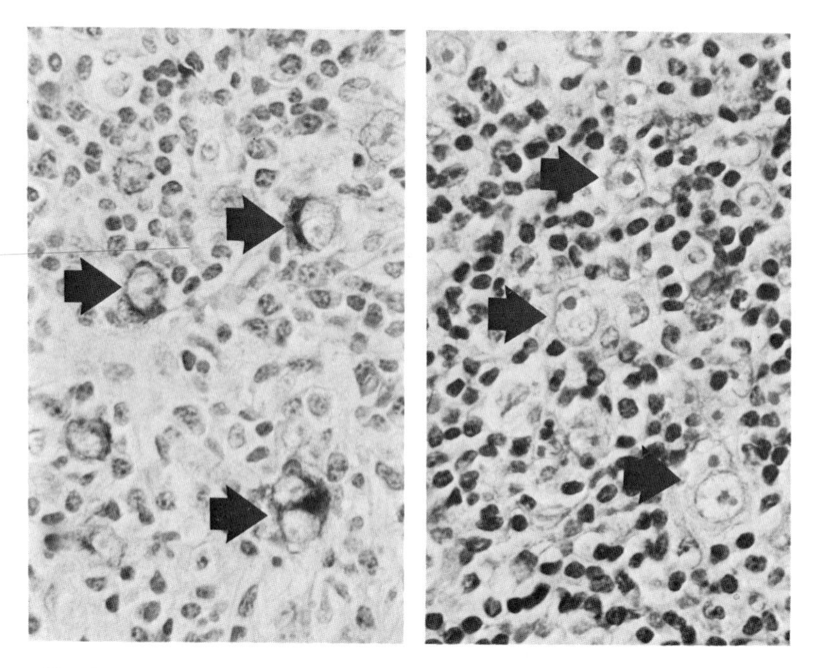

Fig. 9.2. Indirect immunoperoxidase detection of epithelial membrane antigen in nasopharyngeal carcinoma in the left of these adjacent sections. The right section is stained with hematoxylin–eosin. The arrows indicate the individual tumor cells (surrounded by lymphocytes and plasma cells) which demonstrate a strong staining for both the membrane and cytoplasm. Courtesy of Dr. J. P. Sloane with permission to reproduce.

5. Malignant Histiocytosis and Histiocytic Medullary Reticulosis

Functionally and morphologically different categories are included in the so-called diffuse histiocytic lymphomas in the Rappoport classification (Rappoport, 1966). This resulted in the proposal of other classifications with different diagnostic histologic criteria with associated (dis)advantages (Nathwani, 1979).

It has become clear that malignant histiocytosis (MH) cells belong to the mononuclear–phagocyte system (Van Furth, 1975). Lysozyme which normally is present in myeloid and monocyte series of cells has been found to be present in MH cells (Taylor, 1976). Carbone *et al.* (1981) investigated a number of cytochemical and immunohistochemical characteristics of MH cells, e.g., they demonstrate diffuse, sometimes strong reactions with acid

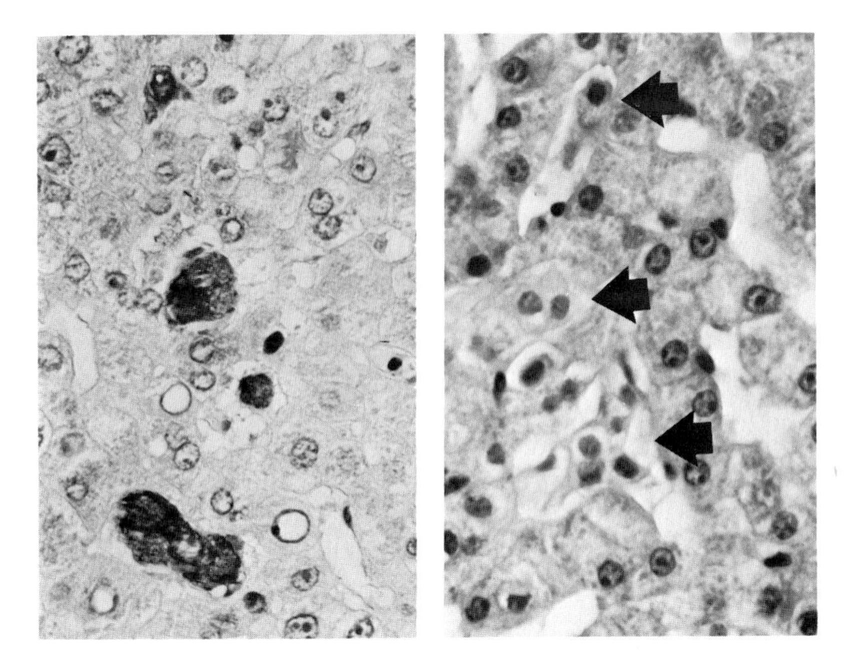

Fig. 9.3. Adjacent sections of metastatic breast carcinoma in liver (single cells and small clumps in sinusoids). The left frame shows a section stained with the immunoperoxidase method for epithelial membrane antigen, whereas the right frame shows the adjacent section stained with hematoxylin–eosin. The arrows indicate some of the tumor cells. Courtesy of Dr. J. P. Sloane with permission to reproduce.

phosphatase (Fig. 9.5). A positive straining for lysozyme was found only in some cases and the intensity of staining varied. This may be due to the loss of detectable enzyme during the course of the disease (Mendelsohn *et al.*, 1980). In contrast, Carbone *et al.* (1981) found that MH cells generally stained positive for both κ and λ light chains (Figs. 9.6 and 9.7). In contrast, staining for immunoglobulins (heavy chains) was often negative. However, caution is warranted since such staining can also be demonstrated in the cytoplasm of cells of carcinematous and sarcomatous proliferations (Banks, 1979).

C. IMMUNOPEROXIDASE APPLICATION IN HISTOLOGY

Intra- and extracellular antigens can be demonstrated on tissue sections or cell smears with the immunoperoxidase method analogous to the immu-

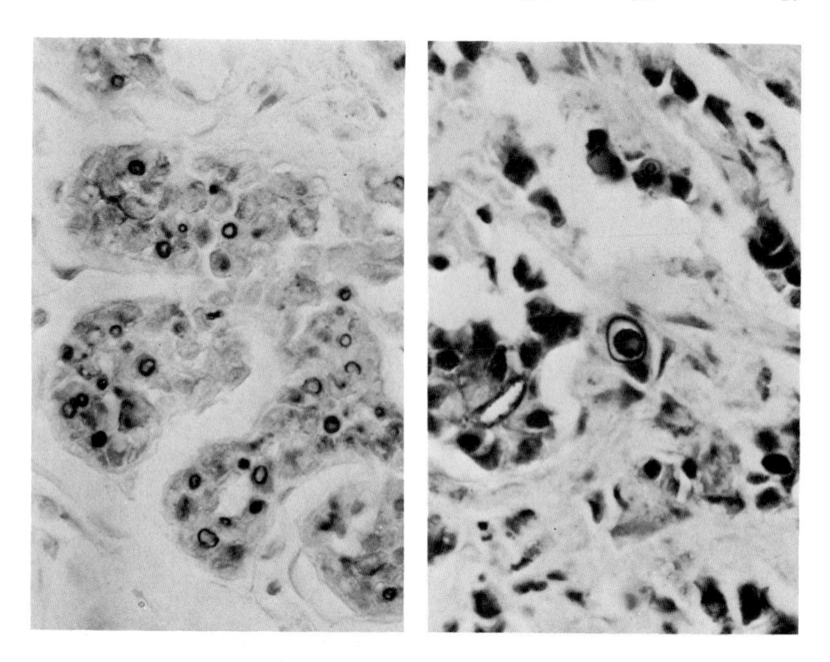

Fig. 9.4. *In situ* lobular carcinoma (left) and infiltrating ductal carcinoma (right) of breast. The lobular carcinoma shows intracytoplasmic lumina, whereas in the infiltrating ductal carcinoma a large target-like intracytoplasmic lumen has pushed the nucleus to the periphery of the cell. Courtesy of Dr. J. P. Sloane with permission to reproduce.

nofluorescence method and ELISA procedures (Kurstak *et al.,* 1977, 1984a,b; Kuhlmann, 1984). Cell surface antigens which are present at low concentrations (e.g., immunoglobulins) are generally difficult to detect, even when unfixed cryostat sections are used.

The main advantage of the immunoperoxidase over the immunofluorescence method in histology is the improved morphologic detail. The combination of suitable fixation procedures, paraffin embedding, and a host of histochemical counterstains provides detailed cytologic detail both of antigen-positive and antigen-negative cells (Taylor and Burns, 1974). Moreover, ultrastructural studies can be undertaken (Kurstak *et al.,* 1977).

Not surprisingly, immunoperoxidase techniques are widely used in histology. It is not our purpose to give an overview of all of these applications, but some studies are chosen which exemplify the usefulness of these techniques.

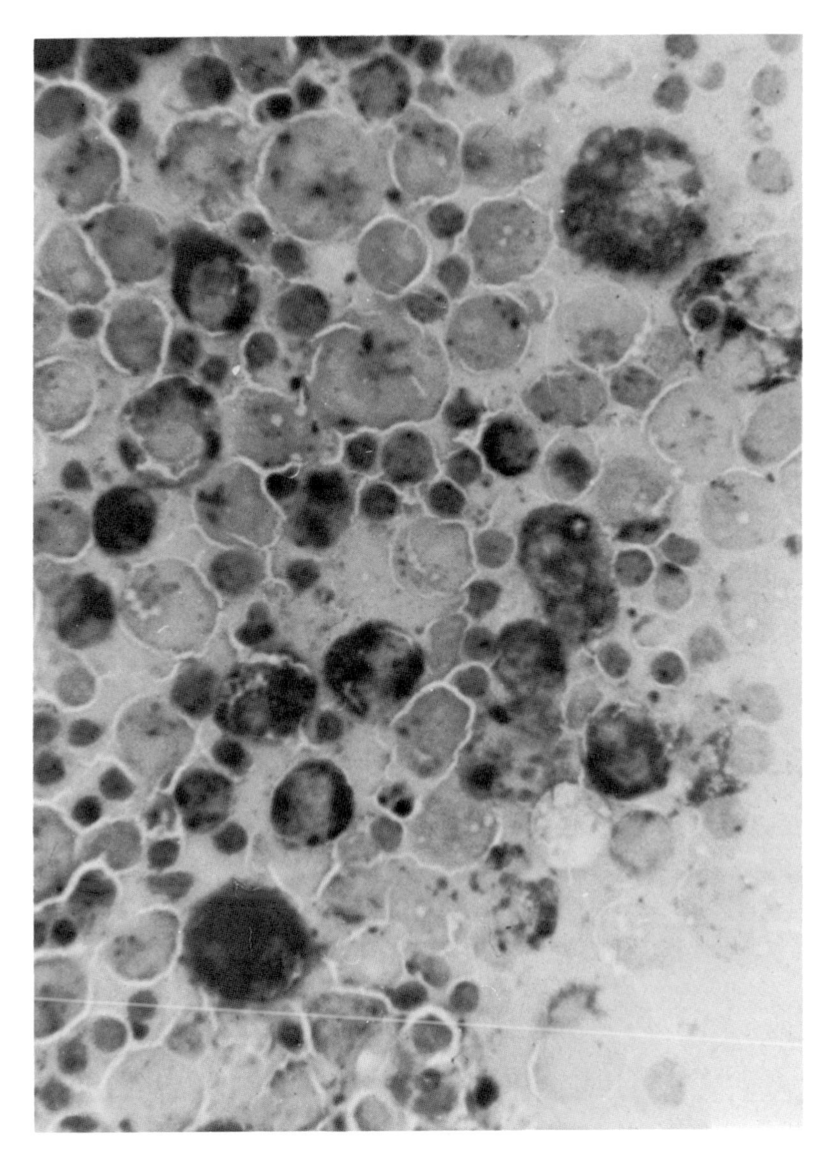

Fig. 9.5. Acid phosphatase staining of malignant histiocytosis cells. Courtesy of Dr. Christian Micheau with permission to reproduce.

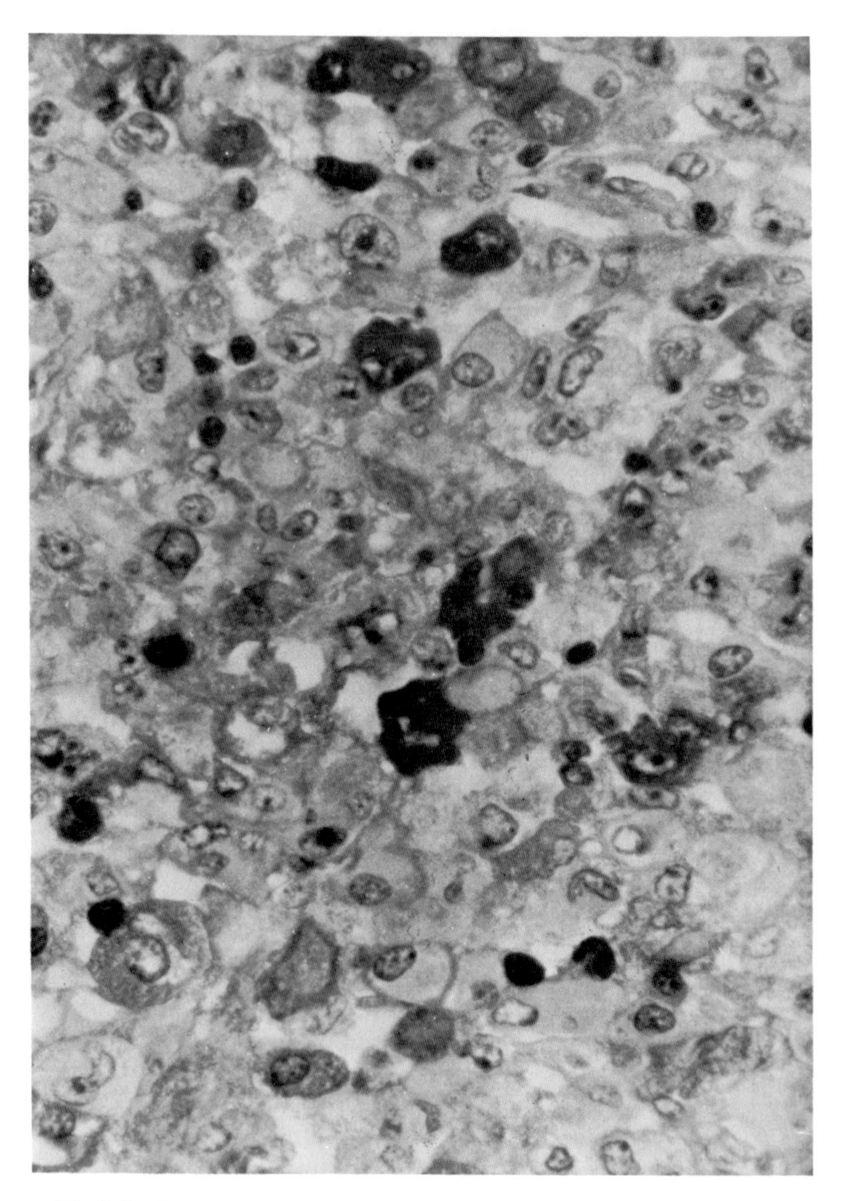

Fig. 9.6. Immunoperoxidase staining (PAP method) for the detection of the κ chain in malignant histiocytosis cells. Courtesy of Dr. Christian Micheau with permission to reproduce.

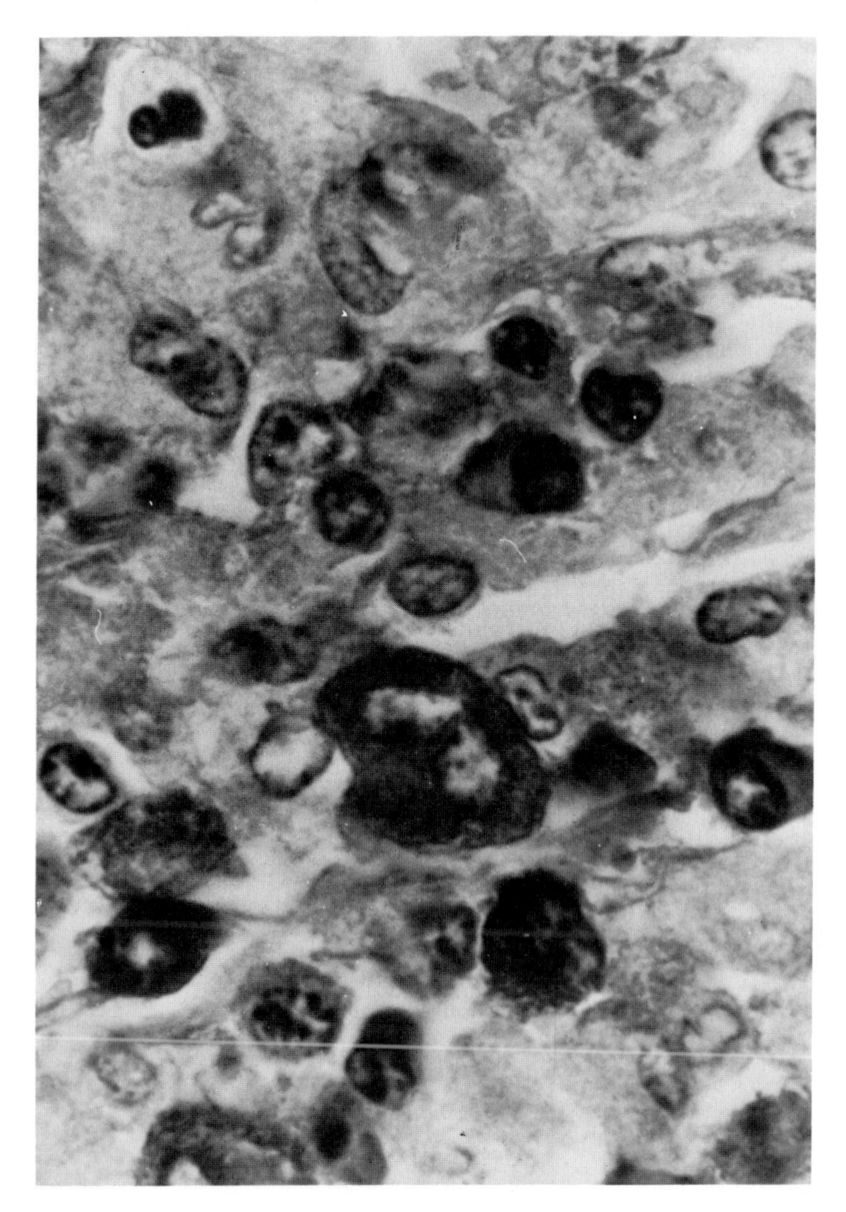

Fig. 9.7. Immunoperoxidase staining by the PAP method for the detection of λ chain in tumorous cells of malignant histiocytosis. Courtesy of Dr. Christian Micheau with permission to reproduce.

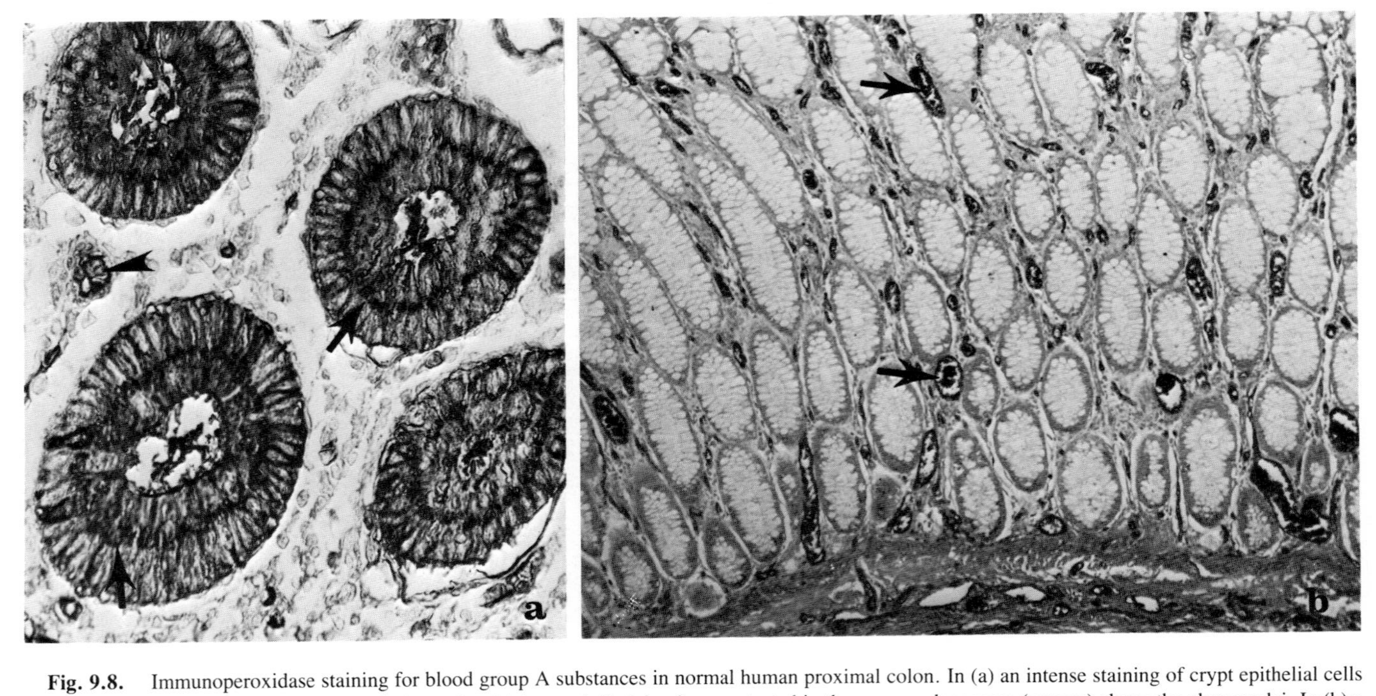

Fig. 9.8. Immunoperoxidase staining for blood group A substances in normal human proximal colon. In (a) an intense staining of crypt epithelial cells and small capilaries with the lamina propria (arrowheads) is noted. Staining is accentuated in the supranuclear areas (arrows) above the clear nuclei. In (b) a higher magnification shows the clear nuclear images and variable staining of mucin vacuoles (arrowheads). Courtesy of Dr. G. Mendelsohn with permission to reproduce.

TABLE 9.4
Localization of A and B Blood Group Substances[a]

1. Deparaffinize sections in xylol and hydrate through graded solutions of alcohol
2. Rinse in 0.05 *M* Tris–HCl buffered saline (TBS, pH 7.6) for 5 min
3. Block endogenous peroxidase activity by rinsing for 30 min in methanol–0.3% H_2O_2
4. Rinse for 15 min in TBS
5. Incubate in 1:20 diluted normal goat serum (Cappel Lab.) for 10 min
6. Rinse in TBS for 10 min
7. Incubate with primary antibody (1:25 diluted) for 60 min; rinse 3 times for 10 min with TBS
8. Rinse for 30 min in TBS
9. Incubate with rabbit anti-primary antibody (diluted 1:40) for 30 min; rinse 3 times for 10 min with TBS
10. Incubate with goat anti-rabbit antibody for 30 min; rinse 3 times for 10 min with TBS
11. Incubate with PAP complex (prepared from rabbit antibody) diluted 1:100 for 30 min
12. Rinse 3 times for 10 min with TBS and flood slides with staining solution (0.08% DAB and 0.01% H_2O_2) and incubate for 5–10 min
13. Wash in tap water (5 min)
14. Counterstain with light green (Fisher), dehydrate, clear in xylol, and mount slides

[a] According to Wiley *et al.* (1981).

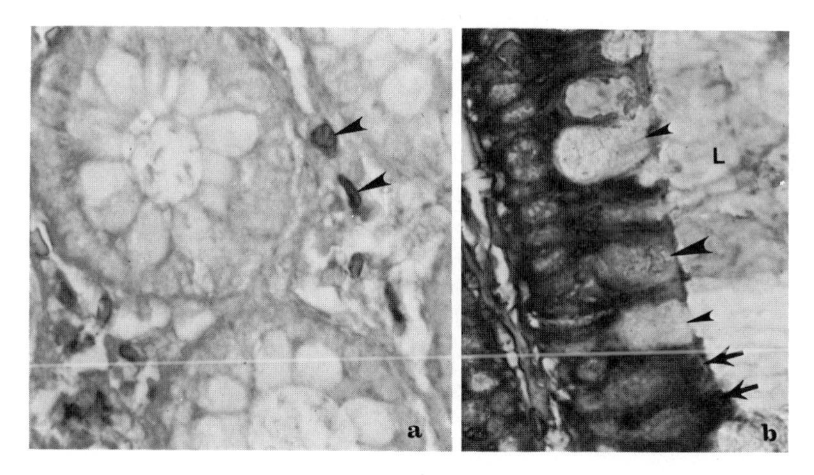

Fig. 9.9. Immunoperoxidase staining for blood group B substances in sections of distal colon. The blood group substances are present only within mucosal capillaries [arrows in (a)] but not within glandular epithelium. A higher magnification (b) of this portion of crypt epithelium confirms that only capillaries are stained (arrowheads; L-luminar mucin). Courtesy of Dr. G. Mendelsohn with permission to reproduce.

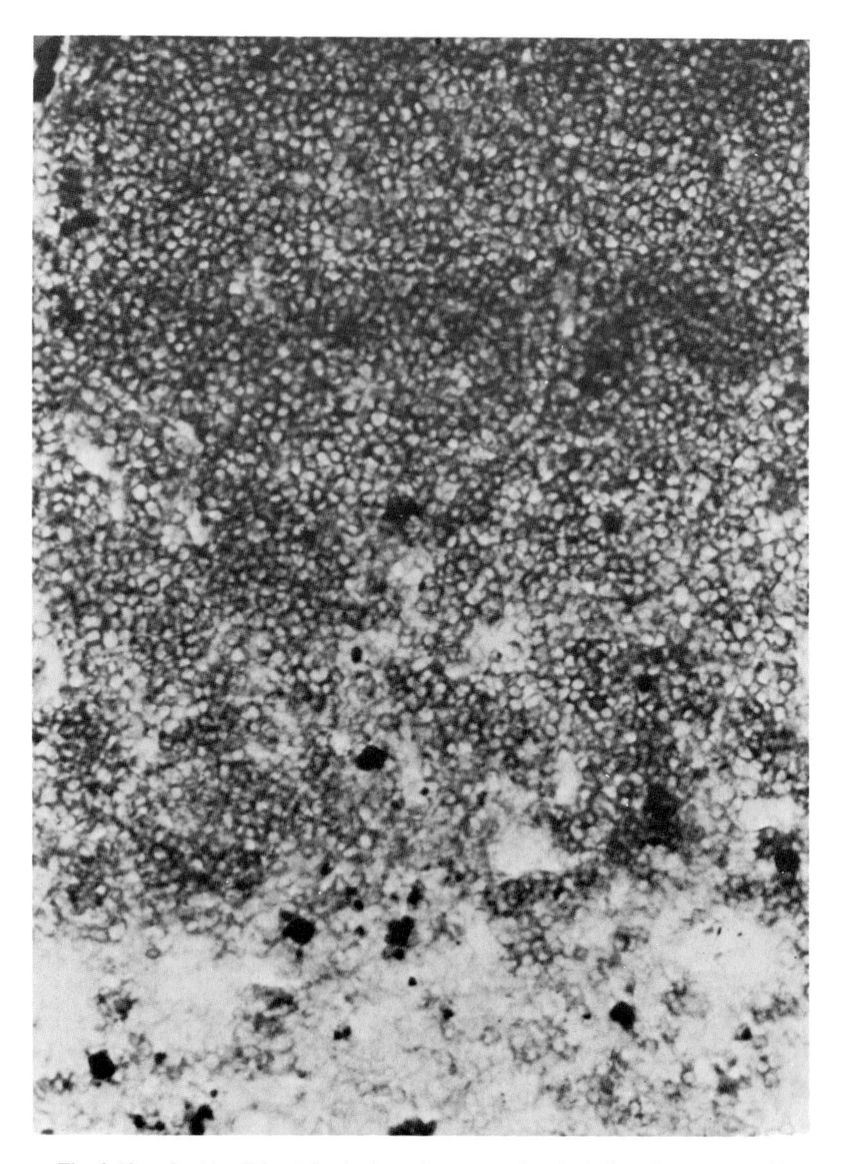

Fig. 9.10. Positive T6 staining in the typic cortex using the indirect immunoperoxidase method. The bottom part of the field shows progressively weakened staining toward the thymic medulla as the medullary thymocytes express the T6 antigen weakly or not at all. Courtesy of Dr. Euan M. McMillan with permission to reproduce.

The blood group substances that determine the blood type are present not only on the erythrocytes but also on the surface of endothelial and epithelial cells of the respiratory, gastrointestinal, and genitourinary tracts and some exocrine organs. Wiley *et al.* (1981) observed the strongest staining in the cecum and the proximal colon, with a progressive reduction of staining in the distal colon and almost complete absence in the rectosigmoid area. Strong staining is obtained on the surfaces of endothelial cells of small blood vessels and serves as a useful internal positive control. The staining was performed as given in Table 9.4. Blood group A substances (Fig. 9.8) were, when present, shown on the columnar cells, on luminal surfaces, and within the cell cytoplasm. The cytoplasmic staining was often heavily concentrated in the supranuclear region of the columnar cells. Mucin vacuole staining varied. Distol colon staining is present only within the mucosal capillaries, as shown with a blood group B patient (Fig. 9.9).

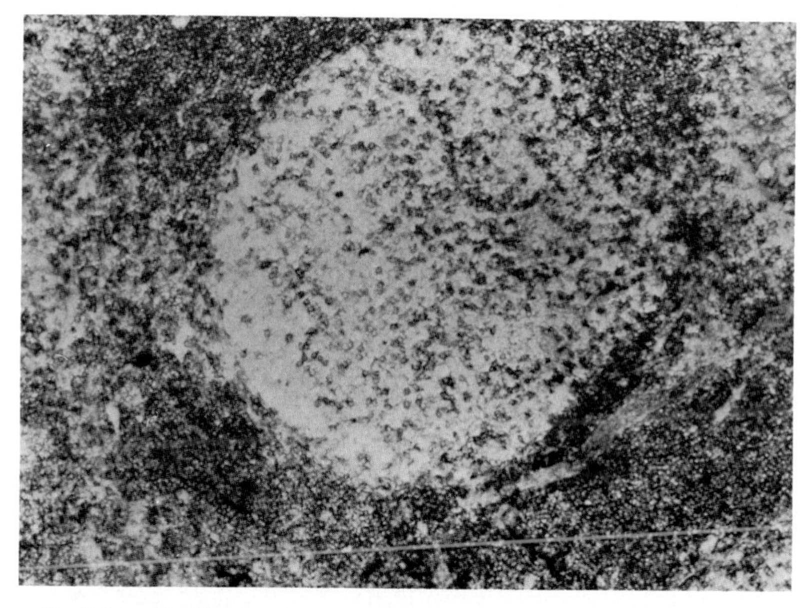

Fig. 9.11. Identification of T lymphocytes and T subsets in human tonsil by the indirect immunoperoxidase technique. The figure illustrates a dense staining of the interfollicular area when using monoclonal Leu 1 (specificity for thymocytes, peripheral blood T cells, and T-dependent areas of lymph node and spleen). Within the lymphoid follicle positively stained cells can also be detected. Courtesy of Dr. Euan M. McMillan with permission to reproduce.

Background staining can be reduced significantly by treatment with Tween 20 at 0.05–2% (Juhl *et al.*, 1984).

The demonstration of OKT 6-positive cells in the human thymus shows the importance of proper fixation (McMillan *et al.*, 1982). The basic procedure is essentially as described in Table 9.4. Various fixatives were used and their influence on staining investigated. Acetate staining resulted in a membranous staining (Fig. 9.10), as did 3% paraformaldehyde in PBS. In contrast, 0.1% glutaraldehyde in PBS or 70% ethanol resulted in a variable staining. No fixation resulted in a poor localization of staining, with poor contrast between membrane staining and background.

The topographic identification of T cells has an important potential since certain disorders may be associated with the depletion of lymphocyte subsets from their normal areas of "heming." The advent of monoclonal antibodies to T cell subsets offers the possibility of studying the distribution of such subsets (McMillan *et al.*, 1981). Leu 1 antibody is specific for thymocytes, peripheral cloog T cells, and T-dependent areas of lymph node and spleen (Figs. 9.11 and 9.12). In contrast anti-Leu 2A is specific

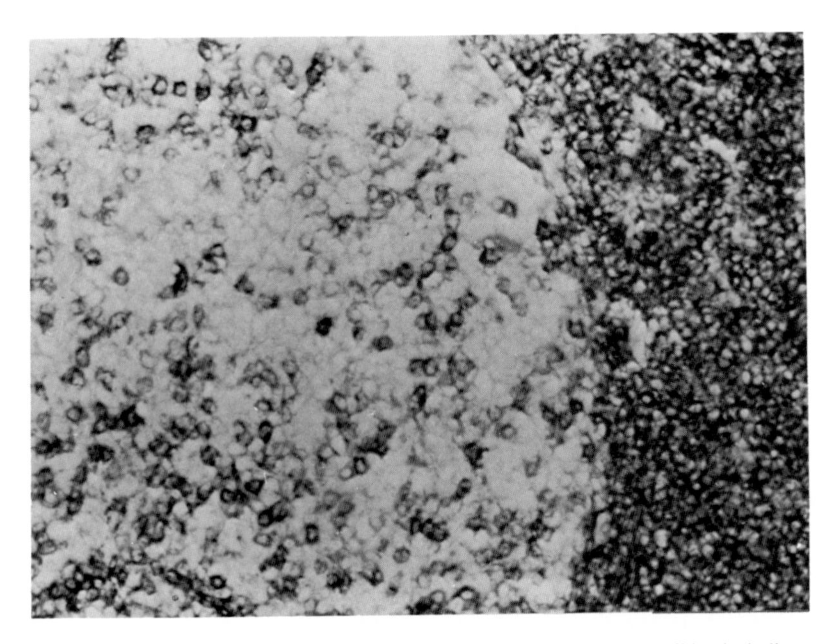

Fig. 9.12. Identification of T lymphocytes and T subsets in human tonsil by the indirect immunoperoxidase technique. This figure shows the primary lymphoid follicle and interfollicular area. Courtesy of Dr. Euan M. McMillan with permission to reproduce.

for cytotonic/suppressor T subsets and anti-Leu 3A for helper/inducer T subsets. Tanaka *et al.* (1984) confirmed that these surface antigens can be shown using acetone-fixed (4°C) and low melting point paraffin wax, resulting in staining similar to fresh-frozen sections. However, it was found advantageous to pretreat the specimen with hyaluronidase.

10

Enzyme Immunoassays for Infectious Diseases

A. INTRODUCTION

The causative agents of infectious diseases are traditionally identified using (i) bioassays, (ii) immunodiagnosis, or (iii) biochemical, biophysical, or microscopic techniques for direct identification. The limitations and merits of these approaches have been discussed by Kurstak *et al.* (1984a,b). In summary, bioassays are generally very sensitive, but are often inadequate (long replication time of organisms, antibiotic treatment of patient interferes with cultivation, and sometimes agents cannot be grown *in vitro*) for rapid diagnosis and treatment. Immunodiagnosis in its various forms will continue to become more important for identification purposes, not only because serological methods have become more refined, but also since the knowledge of the nature of the immune response is increasing rapidly. For example, the paradigm of an IgM primary response and an IgG secondary response seems increasingly oversimplified. However, there will be a shift from the relatively insensitive or complex procedures, such as immunodiffusion, immunoelectrophoresis, agglutination, radioimmunoassay, and immunofluorescence, to enzyme immunoassays. Direct identification will continue to play a role in the diagnosis of infectious diseases, particularly for bacteriology (microscopy), in distinguishing related agents (e.g., electropherograms of rotaviruses), and in the detection of viruses by electron microscopy.

Enzyme immunoassays are rapidly becoming the most important tool in the diagnosis of infectious diseases and increasingly sophisticated formulations of assay systems have been devised. As a general rule, competitive assays are used when specificity is the most important requirement, whereas noncompetitive assays are used when high sensitivity is required. A high specificity is needed when the prevalence of the disease is low, otherwise the majority of positive test results are false positives. In contrast, in the case of diseases with high prevalence, improvements in sensitivity have greater impact on the reliability of the test than improvements in specificity. A numerical example can demonstrate this point. If the prevalence of a disease (i.e., the probability that a positive result is obtained when both specificity and sensitivity are 100%) is 2% and the specificity and sensitivity of the test are both 90%, then a total of 116 positives (18 true positives and 98 false positives) would theoretically be obtained of which only 15.5% are true positives. Increasing specificity will rapidly decrease the number of false positives and thus improve greatly the reliability of the results. For example, increasing the specificity to 95% while keeping the sensitivity at 90% almost doubles the reliability (to 26.9%); conversely, when the sensitivity is increased to 95% but the specificity is kept at 90% the fraction of true positives among the positive test results increases only very slightly, from 15.5 to 16.2%. However, with high prevalence (e.g., 80%) completely different results emerge. When both specificity and sensitivity are 90%, the reliability of positive results is quite good (97.3%), but the reliability of the negative results is relatively poor (69.2%). Thus improvement in the reliability of negative results at high prevalence requires improvement in sensitivity, whereas increasing the specificity has relatively less impact on the reliability of the results. The choice of the enzyme immunoassay in such situations is thus clearly a noncompetitive system with additional steps (enzyme–antibody complexes such as PAP or avidin–biotin systems) for the reasons discussed above. These considerations are of utmost importance but are often overlooked.

Progress in enzyme immunoassays and their applications to infectious disease diagnosis has been achieved at several levels: (i) improvement in the quality and standardization of reagents due to developments in the hybridoma technique and DNA recombinant technology, (ii) more efficient conjugation procedures which minimize loss of enzyme or antibody activities, and (iii) the design of sophisticated assay systems (Kurstak, 1985; Kurstak *et al.*, 1986).

B. ENZYME IMMUNOHISTOCHEMISTRY OF INFECTED CELLS AND TISSUE FROM CULTURES

It is possible to maintain cells or parts of organs after removal from the body. Cell and tissue cultures are frequently used in a variety of disciplines in which enzyme immunohistochemistry has proved a very important adjunct. These studies are quite similar to those with freshly dissected tissue (fixation and/or freezing, sectioning, immune detection, enzyme action). An important exception to this procedure is that sectioning is not required for cells grown *in vitro* as a monolayer. However, permeabilization of the cell membrane is then necessary for intracellular antigens.

Jessen (1983) raised two general cautionary points when using cultures. First, cells may often lose or retain a certain degree of differentiation. For example, Mirsky *et al.* (1980) observed that the rat neural antigen-1 (RAN-1) is barely detectable on Schwann cells prior to culturing, whereas in culture an unambiguous staining of the peripheral glia is obtained. Extrapolation from *in vitro* studies to *in vivo* conditions may, therefore, be unwarranted. Second, high density of cells *in vitro* (e.g., confluence) should be avoided since this leads to a decrease in the quality and reproducibility of the staining. Cells are often best stained at the periphery of the cell layer or at the border of the outgrowth zone.

Infection of cells grown *in vitro* allows the study of the replication of infectious agents. Caution is warranted in extrapolating such findings to *in vivo* systems. Many interacting systems (e.g., immune system) are absent *in vitro*. Nevertheless, such studies show the cell potentials (e.g., *in vitro* transformation).

C. THE USE OF POLYCLONAL ANTISERA OR MONOCLONAL ANTIBODIES IN ENZYME IMMUNOASSAYS FOR INFECTIOUS PATHOGENS

A most important phenomenon among the infectious pathogens is that they may be closely related, as is the case for herpes simplex viruses (HSV-1, HSV-2), or they may be subjected to an almost continuous "antigenic drift" (influenza viruses). The heterogeneity and variability of hyperimmune serum are extremely important characteristics for immunoassays of infectious agents. If two viruses, e.g., HSV-1 and HSV-2, are assayed with a monoclonal antibody to a common determinant, the assay will be unsuccessful in distinguishing the two viruses. However, polyclonal antisera reflect a multitude of individual activities, some of which will be directed to both viruses while others will not. This results in an improved specificity of the polyclonal antiserum (indicated in literature as "specificity bonus"). At the other extreme, however, monoclonal antibodies may be directed to type-specific epitopes on the two viruses resulting in a clear distinction in the assays.

Monospecificity of monoclonal antibodies can, at times, be a disadvantage, and panels of monoclonal antibodies are desired. The disadvantage can be due to the inability to distinguish between agents with cross-reactive determinants, to antigenic drift, or to a loss of avidity (affinity bonus; Ehrlich *et al.*, 1982). Moreover, monoclonal antibodies often fail to precipitate antigens, a disadvantage in procedures that depend on the characterization or isolation of immune complexes. This problem can be overcome by a judicious choice of antibody mixture.

Monoclonal antibodies may be unusually susceptible to denaturation or behave atypically for IgG or IgM molecules. They precipitate often at or near their p*I* which is generally just above pH 7.0. A given monoclonal antibody is of a certain isotype and may not possess certain biological activity, such as complement fixation, making it impossible to use it in complement-based enzyme immunoassays.

D. APPLICATIONS OF ENZYME IMMUNOASSAYS IN PARASITOLOGY

1. Introduction

About 70% of the world's population is affected by infectious diseases, often protozoan or helminthic (Higashi, 1984). Populations with parasitic diseases are usually neglected since they are primarily in developing countries which are constrained due to limited facilities and trained manpower. These economic restrictions combined with an environment ideal for parasite transmission (tropical or subtropical regions) results in widespread parasitic diseases. The simplicity and low cost of enzyme immunoassays are, therefore, real boosts for the diagnosis of many protozoan and helminthic diseases which tend to be the last to benefit from new serological methods. In fact, parasitology has had a prominent position in the development of enzyme immunoassays. Isotope assays are less feasible in developing countries, and the World Health Organization is introducing these nonisotope alternatives with great success.

The usual method for identifying parasitic infections is to examine blood or fecal samples for parasitic eggs, cysts, or larvae by microscopy. However, this method fails or is impractical when extravascular or extraintestinal infections are involved or in cases of epidemics when large numbers of samples have to be processed. In the epidemiological context, enzyme immunoassays are of enormous use in parasitic disease diagnosis.

Infections often are latent in endemic regions and only a minority among those infected may show a clinical morbidity. Moreover, the population may have been previously infected and have multiple infections. Enzyme immunoassays may be of considerable help in these seroepidemiological surveys and also provide a very useful adjunct to traditional methods of identifying parasites in individual patients.

Repeated infection with a certain parasite leads to elevated levels of antibodies in the IgM, IgG, and IgE (in particular helminths) classes, with wide specificities and extensive cross-reactivities. Most important is, therefore, the isolation of specific antigens in order to design assays of high specificities (Schiller, 1967). Simultaneous progress in *in vitro* cultivation, hybridoma techniques, and new separation techniques (e.g., HPLC) has led to significant advances (Mitchell and Anders, 1982).

Very few of the immunodiagnostic tests have the sensitivity or specificity to replace the traditional methods of direct identification. Cross-reactivity is particularly extensive among groups of helminths (cestodes, trematodes, and nematodes). Polyparasitism (simultaneous infection with several parasites) is frequently observed in many regions (Buck *et al.*, 1978), complicating diagnosis or yielding incomplete results.

Enzyme immunoassays are and will be very useful in the follow-up of mass chemotherapy campaigns and environmental control strategies (Higashi, 1984). Particular problems which may arise are (i) false negatives due to the presence of circulating immune complexes, and (ii) false negatives due to the frequent ability of parasites to induce immunosuppression or to evade the immune responses of the host by antigenic masking.

2. Helminthic Infections

Clinical morbidity is often absent in cases of helminthic infections. In fact, morbidity is dependent on the exposure rates (worms do not replicate in humans), and most inhabitants of endemic regions, by adulthood, have been exposed at least once to each parasite. Immunoassays must also have the ability of overcoming extensive cross-reactivity, making it difficult to identify recent infections from previous infections, resulting in a considerable number of false positives (Pifer *et al.*, 1978; Schroeder, 1985).

a. Schistosomiasis

Schistosomiasis is an important widespread tropical disease of man (about 300 million afflicted). The most important parasites of this group are *Schistosoma japonicum* and *Schistosoma haematobium*. They are identified by the species-characteristic eggs in the feces (*S. mansoni* and *S. japonicum*) or in the urine. *S. haematobium* and egg count intensities may be correlated to clinical morbidity.

None of a variety of serological tests is completely satisfactory. The circumoval precipitation test (COPT) in which whole eggs are incubated with serum results in positive cases in precipitates from the surface of the egg shell (Oliver-Gonzalez, 1954). It has 100% sensitivity if the egg count is over 10/g of feces (Ruix-Tiben *et al.*, 1979); results are negative after therapy. Immunoassays may be a significant aid in stages of the infection at which egg excretion becomes scant.

Soluble egg antigens have been fractionated (Mott and Dixon, 1982) and monoclonal antibodies have been raised against them. *S. japonicum* egg antigen-directed monoclonal antibodies have shown great promise in a competitive ELISA test (Mitchell *et al.*, 1983). Crude soluble egg antigens are not very appropriate since they demonstrate cross-reactivity with fascioliasis, hydatidosis, cysticercosis, and trichinosis (Hillyer and Gomez de Rios, 1979). The major antigen in the COPT reaction (Hillyer and Pelley, 1980) had less than 73% specificity or sensitivity in radioimmunoassays, but other fractions of egg antigen have enhanced sensitivity and specificity in ELISA (McLaren *et al.*, 1981).

The differentiation of chronic from acute infections has been attempted and was successful with a proteoglycan from adult worms (Nash *et al.*, 1983). Most patients with active infections have antibodies to the antigen (Kelsoe and Weller, 1978), however, most circulating schistosome antigens were probably in immune complexes.

b. Hydatidosis

Echinococcus granulosis and *Echinococcus multilocularis* (canine tapeworms) eggs ingested by humans develop larvae which migrate to viscera (heart, liver, lungs, brain) where they grow into cystic forms. Morbidity depends on the number and sites of these cysts.

Most of the initial ELISA suffered from a lack of specificity (Ambroise-Thomas and Desgeorges, 1980). However, Capron *et al.* (1970) isolated an antigen (called arc 5) to which antibodies were almost always present in sera from patients (sample of 400). The enzyme-linked electroimmunodiffusion assay using this antigen showed a sensitivity of 88% (300 sera) and a specificity of 100% (1700 sera) (Pinon *et al.*, 1979).

Antigen 5 is not species specific in that it is also present in other tapeworms (Varela-Diaz *et al.*, 1975), and some patients (perhaps 5%) with cysticercosis may also have antibodies to this antigen (Schantz *et al.*, 1980). This does not, however, decrease the diagnostic value of the detection of this antigen since the geographical distribution of these parasites is chiefly separated and clinical differentiation is possible. As with schistosome antigens, circulating immune complexes may be the cause of false negatives (Richard-Lenoble *et al.*, 1978). The strain variation in different regions (Thompson and Kumaratilake, 1982) and the site and number of cysts (least detected in lungs and brains) also contribute to false negative reactions.

c. Cysticercosis

Ingested embryonated eggs of *Taenia solium* hatch in the intestine and larvae enter the circulation. They develop into cysts and may cause severe morbidity depending on the site. False negatives can be quite high with neurocysticercosis (Flisser *et al.*, 1979) probably due to the poor humoral immunoresponse. However, recent advances in the enzyme immunoassay improved detectability to about 75% (Diwan *et al.*, 1982; Espinoza *et al.*, 1982), particularly in antibody assays of cerebrospinal fluid. Improvements in these assays are still required for the detection of this important antigen.

d. Onchocerciasis

Onchocerca volvulus causes ocular complications and skin diseases depending on the worm burden. These diseases occur primarily in Latin American and Africa. In the latter continent several filarial parasites may infect humans so that cross-reactivity is a problem. However, Latin America, particularly Guatemala and Mexico, is usually free of other filarial parasites making immunoassays more specific and useful (Higashi, 1984). Cross-reactions were observed with sera from patients with loiasis or hydatid diseases (Ambroise-Thomas *et al.*, 1980).

Des Moutis *et al.* (1983) produced a monoclonal antibody that readily reacted with circulating antigens. The number of false negatives was, nevertheless, quite high and the amount of circulating antigen was not related to microfilarial densities.

e. *Trichinosis*

One of the infections of domestic animals which should be prevented from reaching man is the helminth *Trichinella spiralis* which frequently infects pigs. The prevalence in man is much higher (about 2% in the United States) than the annually reported cases (100 in the United States). Ruitenberg and co-workers recognized early the value of enzyme immunoassay for mass screening (Ruitenberg and van Knapen, 1977) though much of the early work was done with sera from humans (Engvall and Ljungstrom, 1975). Gamble and Graham (1984) isolated two proteins with MWs of 53,000 and 49,000, respectively, by means of affinity chromatography using monoclonal antibodies. These antigens reacted only with sera from infected swine, and with ELISA as little as 1 larva per 100 g muscle could be detected.

f. *Toxocariasis*

Toxocara canis may cause severe diseases in young children in developed countries (visceral larva migrans and ocular toxocariasis).

Larval extracts have been used to design ELISA tests of reasonably high sensitivity and specificity (Glickman *et al.*, 1979; de Savigny, 1975). An elegant method, developed by de Savigny *et al.* (1979), is the use of secretory antigens. Van Knapen *et al.* (1982) compared the use of secretory larval antigens with that of somatic adult antigens.

3. Protozoan Infections

a. *Malaria*

Over 1 billion people live in endemic regions (Bruce-Chwatt, 1979), and almost half of them live in areas in which transmission is unchecked. Resistance of mosquito vectors to insecticides and the development of drug resistance by *Plasmodium falciparum* pose major obstacles to control measures (Wyler, 1983).

Tests for the parasite should have high species specificity since *P. falciparum* causes almost all of the acute mortality cases. For this purpose major polypeptides have been isolated (Da Silva *et al.*, 1983). Monoclonal antibodies to these polypeptides were effective in neutralization of the parasite for *in vitro* growth. Kemp *et al.* (1983) cloned some genes from *P. falciparum* coding for immunogenic proteins into *Escherichia coli*. This could provide large amounts of antigens for the assays. However, Schofield *et al.* (1982) demonstrated with monoclonal antibodies that differences in reactivity exist among isolates of the parasite from different regions. The ELISA developed to detect *P. falciparum* makes it possible to detect as few as 8 parasites per 10^6 erythrocytes (Mackey *et al.*, 1982). The fraction of false negatives increases, however, when blood contains high antimalarial antibody titers, thus lowering the predictive values.

b. *Trypanosomiasis*

Trypanosoma brucei gambiense causes a chronic infection, whereas *Trypanosoma brucei rhodiensiense* causes an acute disease. Infections with African trypanosomes (sleeping disease) result in high IgM levels with low antibody concentrations, probably since they seem to be strong polyclonal B cell activators.

A problem with the African trypanosomes is the antigenic drift resulting in cyclical antigenic variation. ELISA tests are reasonably sensitive and specific (Ruitenberg and Buys, 1977; Picq *et al.*, 1979), particularly if common variant antigens are used (Vervoort *et al.*, 1978).

Chagas' disease (American trypanomiasis caused by *Trypanosoma cruzi*) is widespread in Latin America (10–15 million; Brener, 1982). Voller *et al.* (1975) developed an ELISA with excellent sensitivity but with a high fraction of false positives (Spencer *et al.*, 1980), particularly with sera from patients with leishmaniasis (Guimaraes *et al.*, 1981). The specificity of this test should therefore be improved.

A very interesting approach has been taken by Crane and Dvorak (1980) who fused *T. cruzi* with murine myeloma cells and obtained hybrids which produced protozoan antigens.

c. *Amebiasis*

Immunoassays for *Entamoeba histolytica* are important for cases of extraintestinal disease since the parasites usually cannot be detected in the stool. ELISA has been employed and has been useful (Bos *et al.*, 1980), particularly in capture assays (Pillai and Mohimen, 1982) or as a kinetic-dependent ELISA (Matthews *et al.*, 1984). The latter requires, for rapid processing, a computer-controlled microplate reader, a drawback in developing countries.

d. Giardiasis

Giardiasis is the most often identified intestinal parasite in the United States. An ELISA developed to detect IgG antibody to *Giardia lamblia* (Smith *et al.*, 1981) was able to detect 80% of 59 proved cases but had 12% false positives.

e. Pneumocystosis

Pneumocystis carinii is an important causative agent of pneumonitis, chiefly in the immunocompromised, children, or persons with malignancies or AIDS (Hughes, 1978; Gottlieb *et al.*, 1981). ELISA for this ubiquitous parasite is not yet well developed (Maddison *et al.*, 1982; Leggiadro *et al.*, 1982), though an urgent need for it exists.

f. Toxoplamosis

Toxoplasma gondii is universally distributed, and often causes clinically inapparent infections. Early tests to detect this parasite or to establish serum titers using enzyme-labeled antibodies were performed with the immunoperoxidase technique (Kurstak *et al.*, 1973) similar to the widely used indirect immunofluorescence antibody test (Fig. 10.1). Subsequent studies used antigen immobilized on microtiter plates (Voller *et al.*, 1976). Class capture assays have been successful for the detection of IgM antibodies to *T. gondii* (Naot and Remington, 1980), thereby avoiding nonspecific reactions due to antinuclear antibody or rheumatoid factors (Payne *et al.*, 1982). Recently Organon Teknika (Belgium) marketed a "Toxonostika" MicroELISA test for detection of IgM and IgG in serum or plasma to *T. gondii* (Table 10.1) which offers a good performance in sensitivity and specificity.

Antigenemia in acute infections to detect circulating antigens is feasible (Araujo and Remington, 1980), but, interestingly, polyclonal antisera from rabbits were better suited for this purpose than murine monoclonal antibodies (Araujo *et al.*, 1980).

In recent work Pinon *et al.* (1985) described an enzyme-linked immunofiltration assay used to compare infant and maternal antibody profiles in toxoplasmosis. The assay is carried out on a micropore membrane. Their doubly analytical method permits simultaneous study of antibody specificity by immunoprecipitation and characterization of antibody isotypes by immune filtration with enzyme-labeled antibodies. This method used to detect IgM antibodies or to distinguish between transmitted maternal IgG

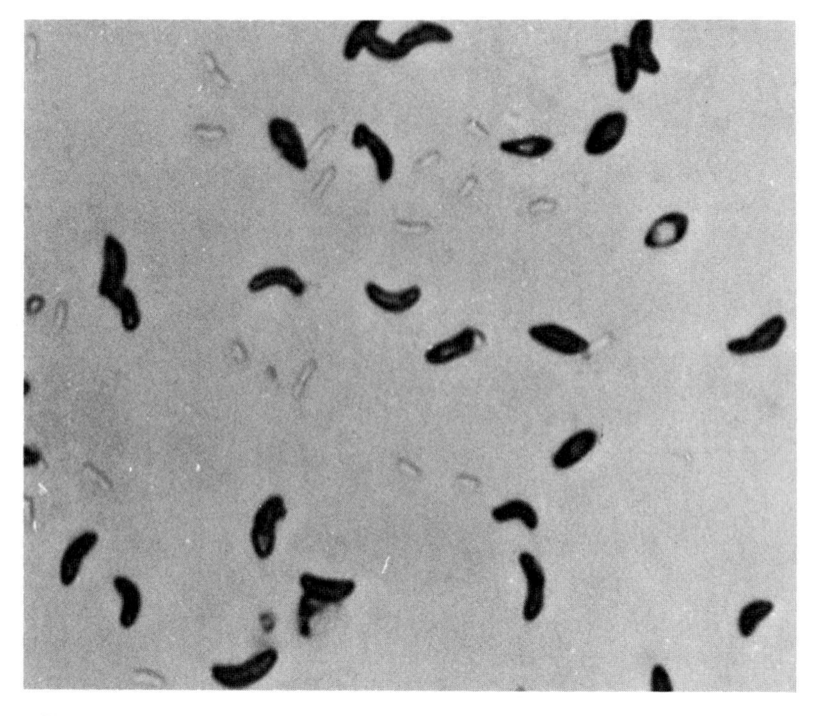

Fig. 10.1. Direct immunoperoxidase staining of *Toxoplasma gondii* parasites. Antibodies to *T. gondii* from a patient were conjugated with peroxidase and used at a dilution of 1:500. The added *Bacillus* bacteria did not react with the conjugate.

and IgG antibodies synthetised by the fetus or neonate makes a diagnosis of congenital toxoplasmosis possible in 85% of the cases during the first days of life. With this method the diagnosis may be made on an average of 5 months earlier than with classical methodology.

The use of monoclonal antibodies in a double-sandwich ELISA was also proposed for detection of IgM antibodies to the major surface protein (P30) of *T. gondii,* especially for the diagnosis of acute acquired toxoplasmosis (Cesbron *et al.*, 1985). The method is based on the capture of serum IgM antibodies, which are revealed indirectly by the addition of a *T. gondii* extract and a β-galactosidase-conjugated anti-P30 monoclonal antibody. An antihuman μ-chain monoclonal antibody was also used for detection of IgM antibodies to *T. gondii* by reverse immunosorbent assay (Pouletty *et al.*, 1985).

TABLE 10.1
MicroELISA Test for Detection of IgM and IgG to *Toxoplasma gondii* in Patients' Serum or Plasma[a]

Reagents

1. Preparation of wash buffer by diluting 1 volume of concentrated buffer with 24 volumes of distilled water (100 ml per strip)
2. The sera must be diluted with diluted buffer (see 1) 1:101 (at least 10 μl of serum with 1 ml of buffer). The controls should be diluted 1:10 with buffer before being brought into the wells or can be pipetted into the wells directly (10 μl of control + 90 μl of buffer)
3. Preparation of IgM conjugate/antigen mixture by adding 0.7 ml of diluted buffer to each ampoule of IgM conjugate and antigen (one for every strip). Mix the two in a ratio of 1:1
4. Preparation of IgG conjugate by adding 1.4 ml of diluted buffer to each ampoule of IgG conjugate (one for every strip) and mix well
5. Preparation of peroxide/substrate buffer by adding 1.0 ml of urea peroxide solution (one U.P. tablet in 10 ml of distilled water) to the vial of substrate buffer
6. Preparation of substrate by adding 0.5 ml of reconstituted peroxide/substrate buffer (see 5) to 5 ml of distilled water and adding 100 μl of TMB solution while mixing. This solution which should be colorless when used is sufficient for four strips

Assay procedure

1. Pipet 100 μl of the corresponding diluted controls (see reagents, 2) into wells A1,A2, A3, and B1,B2, B3 and 100 μl of diluted serum or plasma (2) into the remaining wells of the appropriate MicroElisa strips (codes T1 for IgG and T2 for IgM). Seal with adhesive tape and incube in incubator or a covered water bath with high relative humidity for 60 min at 37°C (for all incubation)
2. Prepare IgM conjugate/antigen (reagent 3) or IgG conjugate (reagent 4). Wash the strips four times with buffer (reagent 1). Add to each well 100 μl of for IgM strips, conjugate/antigen mixture (reagent 3) and for IgG strips, conjugate (reagent 4). Seal with adhesive tape and incubate for 60 min at 37°C
3. Prepare reagent 6. Wash strips four times with buffer (reagent 1). Add 100 μl of substrate (reagent 6) to each well. Incubate (dust free) for 30 min at 20–25°C
4. Stop color reaction by adding 100 μl of 2 mol/1 *N* sulfuric acid
5. Read results photometrically at 450 nm against mean of negative and positive controls

[a] Toxonostika IgM and IgG MicroElisa system from Organon Teknika (Belgium).

E. APPLICATIONS OF EIA IN BACTERIOLOGY, CHLAMYDIOLOGY, AND RICKETTSIOLOGY

1. Bacteria

Enzyme immunoassays in bacteriology are employed, not only for the detection of bacterial antigens or antibodies, but for the determination of extracellular antigen (toxins)–antibody levels and anti-antibiotic antibodies in antibiotic-sensitive patients (Haan *et al.*, 1978, 1979).

For reasons discussed in Chapter 10, Section A, direct detection of the causal agent of a disease is often required, i.e., rapid treatment of the disease is impossible if development of antibodies is awaited. Traditionally, bacteria are identified by culture or by microscopy, procedures which are often more rapid and more precise, respectively, than enzyme immunoassays. Therefore, the latter have not been used as widely as they have been for infectious diseases (virology, parasitology) for which alternative diagnostic methods are not as useful. Nevertheless, enzyme immunoassays are increasingly applied in bacteriology, particularly for cases in which alternatives are less satisfactory. Antibody diagnosis is often essential for retrospective studies, though the immune response to bacteria can be very complex, since cell-mediated immunity is much more important than with smaller antigens (proteins, viruses).

Lipopolysaccharides (LPS) are important in many gram-negative bacteria, and enzyme immunoassays are a prime candidate for their detection and differentiation (Glynn and Ison, 1981). Either the antibody or the antigen can be attached to solid phases. Polysaccharides can be attached to proteins (bridge) by cyanogen halides (Axen *et al.*, 1967) or benzoquinone (Girard and Goichot, 1981).

Stiller and Nielsen (1983) described a method for LPS conjugation which is considerably more efficient than others for the affinity purification of bovine antibodies to *Brucella abortus* LPS (based on the acetonitrile–CNBr method). This method allows large-scale isolation of affinity-purified antibody to LPS, which has a considerable potential for the standardization of serological procedures and for screening and confirmation tests (Butler *et al.*, 1980; Butler, 1981).

Kusama (1983) developed a double-antibody sandwich method for the detection of LPS from the eight most prevalent *Pseudomenas aeruginosa* serotypes. The importance of *P. aeruginosa,* an opportunistic pathogen in nosocomial infections, has been increasingly recognized. Its LPS are specific, highly immunogenic, and responsible for various clinical manifestations. They are released in surrounding medium. Kusama (1983) observed that LPS II has a higher reactivity in enzyme immunoassays than LPS I (fractions I and II are separated by gel filtration). This could be due to differences in avidity since Morrison and Leive (1975) observed similar differences in immunodiffusion studies on *E. coli* 0111:B4 LPS fractions. However, Goldman *et al.* (1982) showed that *E. coli* LPS I has neither lipid A nor an oligosaccharide core. Munford and Hall (1979) improved the detectability of gram-negative bacterial LPS by increasing the solubility of LPS with triethylamine, but Kusama (1983) failed to confirm this with enzyme immunoassays. However, Kusama demonstrated that enzyme immunoassays using LPS II are highly sensitive or specific for the various serotypes of *P. aeruginosa*.

Carlsson *et al.* (1976) screened maternal serum, cord blood, and human milk for antibodies to *E. coli* 0 LPS. Similarly, Gripenberg *et al.* (1979) demonstrated antibodies against *Yersinia enterocolitica* LPS in human serum by enzyme immunoassays. Results quite often differed considerably from those obtained with hemagglutination. The antigen itself could also be detected, but only if 0.5 μg/liter was present.

Members of the genus *Brucella* infect different species of animals, and some members of this genus may be also transmitted to humans, e.g., *B. suis* which normally infects swine, *B. abortus*, which normally infects bovines, and *B. melitensis* and *B. canis*, which normally infect sheep and dog, respectively; *B. ovis* (from sheep) is not known to infect humans. Though *Brucella* is cultivable, brucellosis has a variable incubation time with an abrupt or insidious onset, so that the majority of tests are confirmatory and based on serology, particularly agglutination (McCullough, 1976). Enzyme immunoassays are most useful, especially in veterinary medicine, where carriers can be detected by the examination of milk and serum antibodies (Byrd *et al.*, 1979; Thoen *et al.*, 1983) and so serve in eradication programs. However, ideally the microbial antigen should be detected directly. Perera *et al.* (1983) devised an enzyme immunoassay with a very high detectability for LPS produced by *Brucella*. They were able to detect 100 fg/ml of solubilized crude LPS (8–10 *Brucella* cells). After concentration of the sample (serum or plasma) LPS could not be detected, though 10 brucellae added to a suspension of leukocytes from 100 ml of normal bovine blood was readily measured. The failure to detect LPS in concentrated or undiluted serum may be due to inhibitory factors. Though *B. abortus* is shed in milk, detection is most imperative in pregnant (last trimester) nonlactary animals.

Many diseases require rapid diagnosis for proper treatment. Culture is often too slow or made impossible through the early administration of antibodies (Glynn and Ison, 1981). For example, the bacterial causes of acute meningitis are most often diagnosed by counterimmunoelectrophoresis (Greenwood *et al.*, 1971). However, this technique fails to detect antigen in a substantial number of patients (Scheifele *et al.*, 1981; Thirumoorthi and Dajani, 1979; Collins and Kelly, 1983). However, enzyme immunoassays have at least a 100-fold higher detectability (Drew *et al.*, 1979) and *E. coli* K 100 (antigenically closely related) does not cross-react. On the other hand, Kaplan *et al.* (1983) employed enzyme immunoassays for the detection of capsular antibodies against *H. influenzae* type b. For this purpose, polyribosephosphate (PRP in the capsule of *H. influenzae*) is coupled to poly-L-lysine, since alone it does not adsorb reliably to microtiter plates. Prior adsorption of anti-PRP antibodies to the plastic often results in excessive nonspecific adsorption of test sera. Comparisons with radioimmunoassays (Kaplan *et al.*, 1983; Koskela and Leinonen, 1981) are inconclusive with respect to detectabilities and sensitivities. This is far from surprising since the underlying principles of these methods are different. Most often radioimmunoassays for *H. influenzae* antibodies are based on saturation analysis, whereas the enzyme-mediated method is based on an immunometric assay. Their inherent sensitivity and specificity levels are different, independent of the label used. The most common cause of neonatal sepsis and meningitis is group B *streptococcus* and the mortality for infants with early-onset infections is about 50%. To prevent vertical transmission, a rapid identification of mothers whose infants may develop infection is necessary. Morrow *et al.* (1984) developed a monoclonal sandwich enzyme immunoassay which was able to detect 1 ng/ml of native antigen and is serotype specific, whereas Polin and Kennett (1980) devised a competitive immunoassay, also based on monoclonal antibodies, for the same purpose. Meningococcal meningitis is usually diagnosed by culture. However, the detection of these antigens by enzyme immunoassays compares advantageously to other serological procedures (Sippel and Voller, 1980; Sugasawara *et al.*, 1984), in particular in the latter study when monoclonal antibodies are used (Sugasawara *et al.*, 1984).

Heat-labile enterotoxins (*Vibrio cholerae, E. coli* pathogenic strains) have been detected by conventional enzyme immunoassay to great advantage (Holmgren and Svennerholm, 1973; Svennerholm and Holmgren, 1978, Svennerholm and Wiklund, 1983; Yolken *et al.*, 1977). Beutin *et al.* (1984) modified these methods by using nitrocellulose membranes as the solid phase (detectability 1 ng/ml). About 95% of normal adults have *Clostridium perfringens* in their fecal flora and identification of the organism would be insufficient for diagnosis (Bryan, 1969). *Clostridium* eneterotoxins have, however, been successfully detected by enzyme immunoassays (Lewis *et al.*, 1981; Lyerly *et al.*, 1983; McClane and Strouse, 1984; Jackson *et al.*, 1985). McClane and Strouse (1984) noted that the presence of fecal material apparently decreases the sensitivity of the sandwich assay. It is of interest to note that this material generally has high proteolytic activity, which may cause the desorption of the capture antibody immobilized on the plate (Viscidi *et al.*, 1984).

Some bacteria have a nonspecific affinity for IgG (e.g., *Staphylococcus aureus* may contain high levels of protein A) which may then interfere in enzyme immunoassays. To prevent these undesired reactions, normal serum may be added prior to the tests (Koper *et al.*, 1980). *S. aureus* is important among hospital-acquired infections, and enzyme immunoassay offers advantages over the gel diffusion and counterimmunoelectrophoresis techniques generally used for the detection of *S. aureus* infections (Carruthers *et al.*, 1984). A comparison of four versions of enzyme immunoassays showed that the sandwich design with labeled antibody is most satisfactory (Fey *et al.*, 1984) and has a detectability for staphylococcal enterotoxins of 0.1 ng/ml which is far below clinical relevance. This test kit is now commercially available.

Many of the procedures for the detection of toxins are based on sandwich methods. These methods require that the second antibody layer (i.e., after solid-phase antibody and test sample with toxin) is either conjugated with enzyme or from a different species so that an anti-immunoglobulin would not react nonspecifically with the solid-phase antibody. This may pose problems in many cases since antibodies for, e.g., staphyloccal enterotoxins presently available are all derived from rabbits, prohibiting the use of anti-rabbit IgG, and many laboratories are not in a position to custom-conjugate enterotoxin antibodies. de Jong (1983) devised a simple but elegant method to circumvent this problem. In his test, de Jong immobilized antibody (1000× diluted in carbonate buffer) and applied the test sample which may contain the enterotoxin. de Jong then mixed an 800-fold dilution of enterotoxin antibody and a 50-fold dilution of anti-rabbit IgG alkaline phosphatase conjugate and incubated this for 2 hr at 37°C; normal serum (10-fold dilution) was then added, incubated for 30 min, and added to the plates. The detectability of this test was at least 1 ng enterotoxin per 1 ml, which is the lowest level considered necessary for enterotoxin testing (Reiser *et al.*, 1974).

A complication with endotoxin detection by enzyme immunoassays is the frequent contamination of commercial peroxidase or alkaline phosphatase conjugates with endotoxins (Klipstein *et al.*, 1984). For example, commercially obtained bovine alkaline phosphatase may contain as much as 1 μg endotoxin per ml. The ubiquity of endotoxin in conjugates and the prevalence of antibodies to endotoxin in mammalian sera may significantly alter the values obtained and the reliability of the assays.

Enzyme immunoassays offer a nonculture alternative for the detection of *Neisseria gonorrhoeae* antigen from urogenital swabs, and a commercial kit (Gonozyme) is marketed by Abbott Laboratories (Table 10.2). Gonozyme is a solid-phase enzyme immunoassay to detect gonococcal antigen

TABLE 10.2
Enzyme Immunoassay for the Detection of *Neisseria Gonorrhoeae* in Urogenital Swab Specimens[a]

Reagents[b]

1. Treated beads
2. Anti-rabbit IgG (goat): peroxidase conjugate (minimum concentration 0.1 μg/ml in Tris buffer)
3. Antibody to *N. gonorrhoeae* (minimum concentration of 0.1 μg/ml in Tris buffer)
4. Extract of *N. gonorrhoeae* in phosphate-buffered saline (PBS). Positive control
5. PBS, negative control
6. PBS, Specimen dilution buffer
7. OPD (*o*-phenylebediamine–2 HCl) tablets
8. Diluent for OPD. Citrate-phosphate buffer containing 0.02% hydrogen peroxide (H_2O_2)

Assay procedure

First incubation

1. Pipet 200 μl of controls (3 negative and 1 positive) or specimen into appropriate wells or reaction tray
2. Add one bead to each well containing specimen or control
3. Apply cover seal and gently tap the tray to cover bead with sample and remove any trapped air bubbles
4. Incubate at 37°C for 45 min
5. Remove and discard cover seal. Aspirate the liquid, and wash each bead three times with 4 to 6 ml of distilled or deionized water

Second incubation

6. Pipet 200 μl of antibody to *N. gonorrhoeae* into each reaction well
7. Apply new cover seal and gently tap the tray to remove any trapped air bubbles
8. Incubate at 37°C for 45 min
9. Remove and discard cover seal. Wash each bead three times

Third incubation

10. Pipet 200 μl of antibody to rabbit IgG (goat): peroxidase conjugate into each well
11. Apply new cover seal and gently tap the tray to remove any trapped air bubbles
12. Incubate at 37°C for 45 min
13. Prepare OPD substrate solution (during the last 5–10 min of the incubation)
14. Remove and discard cover seal. Wash each bead three times
15. Remove all excess liquid from the tray by aspiration or blotting

Color development

16. Immediately transfer the beads to properly identified assay tube
17. Pipet 300 μl OPD substrate solution into each tube (substrate must not touch metal) and into two empty tubes (substrate blanks)
18. Cover and incubate for 30 min at room temperature (15–30°C)
19. Add 1 ml of acid to each tube. Agitate to mix

Photometric reading

1. Blank spectrophotometer with substrate blank
2. Determine resorbance at 492 nm or use the Quantum Analyzers

(*continued*)

TABLE 10.2 (*Continued*)

Results

The presence or absence of *N. gonorrheae* is determined by relating the absorbance of the unknown specimen to the cutoff value. The cutoff value is the absorbance of the negative control mean (NC$\bar{X}$) plus factor 0.190. Unknown samples with absorbance values greater than or equal to the cutoff value established with the negative control are to be considered positive for *N. gonorrhoeae*

[a] Gonozyme test according to Abbott Laboratories.
[b] All solutions with antimicrobial agents.

from urethral or endocervical swabs. This test may detect *N. gonorrhoeae* that fail to grow on selective media because of the sensitivity of the bacteria to an antibiotic or because they are not viable. In the Gonozyme test treated beads are incubated with a swab specimen and appropriate controls. The bacteria from the positive swab specimen adsorb to the bead. The bead is incubated with antibody to *N. gonorrhoeae* which reacts with gonococci on the bead. Next, the bead is incubated with antibody–enzyme conjugate containing horseradish peroxidase which reacts with the antigen–antibody complex on the bead. The enzyme reaction after incubation of the washed bead with *o*-phenylenediamine (OPD) containing hydrogen peroxide (H_2O_2) is determined using a spectrophotometer with wavelength set at 492 nm.

Current methods used to identify *N. gonorrhoeae* in endocervical cultures appear to detect less than 90%. Papasian *et al.* (1984) evaluated the commercial enzyme immunoassay kit and found a sensitivity and specificity of about 95–98% for specimens from males and 80–88% from females; it would appear that this kit is less suitable for the detection of *N. gonorrhoeae* in females. This study confirms other reports by Aardoom *et al.* (1982) and Schachter *et al.* (1984) but contradicts a similar evaluation by Danielsson *et al.* (1983). All investigations agree that this assay has high sensitivity and specificity for detection in specimens from males. The wide discrepancies of results for specimens from females may be due to differences in specimen collection, or may, as Papasian *et al.* (1984) pointed out, be due to a lack of immunological specificity. The positive predictive value using this commercial kit was about 25% lower for females and needs improvement. It is possible, however, that the choice of culture as standard may not be optimal since this standard may be false negative, and the commercial kit result may in fact be correct, though studies of Stamm *et al.* (1984) suggest that this is not so. However, this assay could have a major impact on the control of the spreading of gonorrhea if improved for cervical gonorrhea. For symptomatic gonorrhea, the Gram stain is inexpensive and adequate.

In their study of Gonozyme kit, Smeltzer *et al.* (1985) found that the ELISA procedure, which is based on the detection of outer membrane complex of *N. gonorrhoeae,* is an alternative method of gonorrhoeae diagnosis in both men and women. Geiseler *et al.* (1985) pointed out that Gonozyme is comparable with Gram stain for rapid diagnosis of gonorrhea in females.

Enzyme immunoassays, the Western blot, and related techniques were used for the characterization and detection of immunodominant surface-exposed protein antigens of *Treponema pallidum* (Moskophidis and Müller, 1985). Using monoclonal antibodies to these antigens it was possible to diagnose human syphilis infection and to characterize *T. pallidum* protein antigens.

2. Chlamydia and Rickettsiae

The most common sexually transmitted disease is the infection with *Chlamydia trachomatis* (Holmes, 1981; Morisset and Kurstak, 1985). Many sexually transmitted infections and respiratory tract infections are asymptomatic, but they are not benign, and syndromes attributed to this agent are cervitis, urethritis, salpingitis, proctitis, conjunctivitis, pneumonia, and infertility in women (Jones *et al.*, 1982, 1983).

The common method of diagnosis is the isolation of the organism in tissue culture. This method is expensive and the recovery rate is affected by many variables (Reeve *et al.*, 1975). The identification of chlamydial antibodies is not as specific as tissue culture but is a very useful adjunct, particularly for epidemiological studies. Though Schachter *et al.* (1982) suggested that the detection of IgM is the method of choice for the diagnosis of chlamydial pneumonia in infants, Finn *et al.* (1983) found poor correlation of IgM detection by ELISA with current methods, and Mahony *et al.* (1983) observed limited sensitivity of chlamydial IgM by ELISA. In contrast Finn *et al.* (1983) found satisfactory correlations between these techniques and IgG detection by ELISA. Serotype L2 elementary bodies were used as antigens in these studies. Jones *et al.* (1983) compared the reticulate and elementary body antigens in the ELISA detection of antibodies to *Chlamydia trachomatis*. Elementary body antigen from a single immunotype may not detect antibodies to another immunotype, whereas the reticulate body antigens appear group reactive. Nevertheless, Jones *et al.* (1982, 1983) found broad reactivities for both antigens. Less reticulate than elementary body antigens was required to obtain equivalent binding of antibody. Mahony *et al.* (1983) observed, however, that in their ELISA test the few false negatives obtained were of the C and J serotypes which possess little antigenic relatedness with serotype L2. A major serological problem in the diagnosis of genital chlamydial infections remains the relatively high background (Wang *et al.*, 1977; Schachter *et al.*, 1979). Some of the background may be due to autoantibodies (Sarkku *et al.*, 1983) known to occur in chlamydial infections (Andersen and Møller, 1982). The immunoperoxidase method used in our laboratory to detect *C. tramomatis* antigens in baby hamster kidney cells (BHK-21) with patients specimens was satisfactory (Fig. 10.2) and superior to any other histochemical staining. Raymond *et al.* (1985) demonstrated that a good agreement of results were obtained by the ELISA and microimmunofluorescence assay for detection of antibodies to *C. trachomatis*.

An enzyme immunoassay for the detection of *Chlamydia trachomatis* in urogenital swab specimens (Chlamydiazyme diagnostic kit) from Abbott Laboratories based on the same principle as the Gonozyme is widely used. Chlamydiazyme utilizes a solid-phase EIA to detect chlamydial antigen from urethral or endocervical swabs providing results in less than 4 hr.

Serodiagnosis of Rickettsiae proved possible and sensitive enzyme immunoassays have been developed (Dasch *et al.*, 1979; Crum *et al.*, 1980).

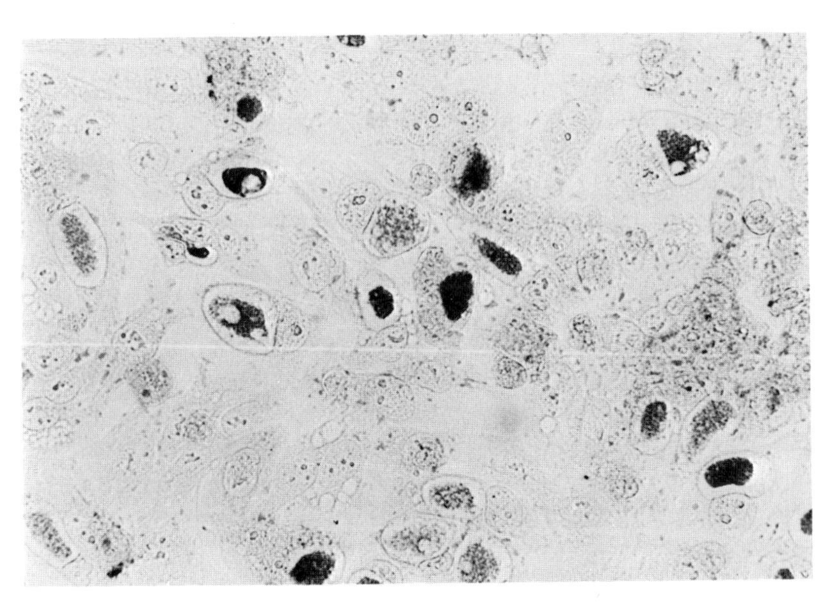

Fig. 10.2. Positive immunoperoxidase staining of antigenic material of *Chlamydia trachomatis* in BHK-21 monolayers infected with patient's specimens.

Dasch *et al.* (1979) adopted the ELISA for the serological diagnosis of infection with the scrub typhus rickettsiae, *Rickettsia tsutsugamushi*. Its strains show an important antigenic diversity and antigens are not easily prepared. Dasch *et al.* (1979) showed that a single antigen could be used for the different strains, though differences in titers are observed, and that ELISA is preferred over immunofluorescence methods. Crum *et al.* (1980) developed a paper ELISA for the same purpose. Dasch (1981) developed a simple procedure for the selective isolation of the protective species-specific protein antigens of *Rickettsia typhi* and *Rickettsia prowazekii* which were satisfactory for ELISA.

Serodiagnosis of scrub typhus (tsutsugamushi fever) also proved possible with the indirect immunoperoxidase method (Fig. 10.3; Yamamoto and Minamishima, 1982). A 10% homogenate of spleens of mice infected with the rickettsiae were injected intraperitoneally to mice, followed by a cyclophosphamide injection. Peritoneal smears were obtained and used to detect antibodies in the sera of patients. The Weil–Felix reaction test is overrated. In fact, half of the fatal cases in Japan in 1980 were negative with this test.

F. ENZYME IMMUNOASSAYS IN VIROLOGY AND FOR THE DIAGNOSIS OF VIRAL DISEASES

1. The Detection of Viral Antigens or Antibodies

The ideal enzyme immunoassay would enable the direct detection of viral antigens and thus offer the possibility of rapid disease management. Unfortunately, this is often not possible (low concentration, different serotypes, viremia at certain stages of the disease).

Serological testing can be helpful in different situations, such as (i) in the case of virus detection but with equivocal interpretation, (ii) in cases in which virus is suspected but the direct detection is difficult or very time consuming (e.g., myocarditis–pericarditis due to coxsackieviruses type 1 through 5 or possibly influenza A or B, rubella due to rubellavirus, central nervous system syndromes due to HSV, mumps, western or eastern equine encephalitis, St. Louis encephalitis and California encephalitis, and perhaps lymphocytic choriomeningitis or Epstein–Barr virus), (iii) in heterophile-negative mononucleosis syndromes (cytomegalovirus, Epstein–Barr virus), and (iv) in situations in which the immune status should be established from single sera (Rubella, hepatitis B, varicella-zoster).

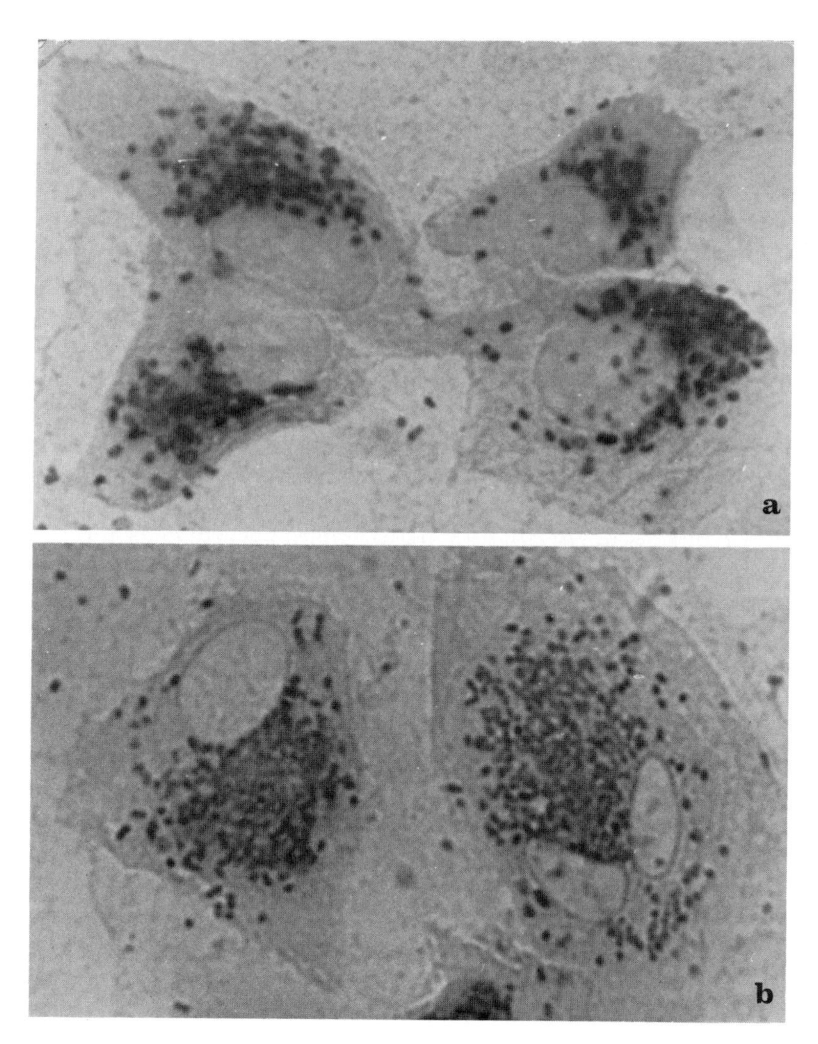

Fig. 10.3. Indirect immunoperoxidase staining of *Rickettsia tsutsugamushi* on cytosmears of peritoneal cells. The sera from the patients were used as primary antibodies. The peroxidase-conjugated goat IgG were directed to either the $F(ab')_2$, as shown in (a), or to the Fc, as shown in (b), fraction of the human IgG. Courtesy of Dr. Yoichi Minamishima and Dr. Seigo Yamamoto with permission to reproduce.

2. Nature of Antigenic Drift of Viral Antigens

Generally, viruses are excellent immunogens and antigens. Many viruses have antigens or epitopes which are specific for their group [group-specific (gs) antigen], e.g., the rotaviruses, though for the latter some so-called pararotaviruses have been discovered which lack this gs antigen.

The rapid replication of viruses and the evolutionary stress due to eradication or protection programs may result in an antigenic drift for some viruses. This is particularly evident for type A influenza viruses (Laver, 1984). Some of the antigenic shifts are massive (major antigenic shift). Suddenly, a major antigenic shift in these viruses may occur by the appearance of viruses with a different hemagglutinin (sometimes neuramidase) due to the complete replacement of the RNA segment coding for this antigen. Twelve distinct hemagglutinin subtypes are recognized by WHO. All of the hemagglutin subtypes exist in viruses from birds or lower mammals. There is also a slow but steady antigenic drift. A large number of these antigenic variants were isolated and characterized with monoclonal antibodies (Gerhard and Webster, 1978). The three-dimensional structure of hemagglutinin has recently been characterized (Wilson *et al.*, 1981). It contains four independent antigenic areas, one of which is buried, at the distant end of the molecule. Analysis of sequence changes of field strains revealed that this drift does not occur at the same position but may be at a neighboring position (Laver, 1984) and that only a single sequence change may be sufficient to totally abolish the affinity of the monoclonal antibody to the hemagglutinin. A similar situation seems to exist for neuramidase (Webster *et al.*, 1982; Schild, 1984).

Unfortunately, relatively little is known about the nature of the cell-mediated immune response and the recognition by T-helper cells of the epitopes on the hemagglutin of the different subtypes.

Investigations on the chemical nature of the viral–antigen drift will be crucial for the development of reliable immunoassays, since detection of the antigen is more critical for diagnosis than the detection of antibody.

3. IgM Determinations

a. *The Usefulness of IgM Determinations in Diagnostic Virology*

In the primary immune response, the antibodies produced are usually of IgM class though it is now recognized that IgG antibodies appear almost simultaneously (Cradock-Watson *et al.*, 1979). The transient nature of the

IgM antibodies in the primary response provided a tool to recognize primary viral infection (Schluederberg, 1965) and has been used as the basis of routine laboratory diagnosis.

In order to distinguish IgM from IgG antibodies several physical methods have been used based on molecular weight or charge differences, such as gel filtration sucrose gradient centrifugation, selective precipitation of IgG, and ion-exchange chromatography. Another method used is based on the observation that the avidity of IgM decreases drastically when the subunits of the IgM molecule are dissociated by treatment with a mercaptan (Banatvala *et al.*, 1967). However, to obtain a fourfold decrease in the titer at least 75% of the total antibodies must be of the IgM class (Meurman, 1983). This method is, therefore, not very sensitive (Forghani *et al.*, 1973).

In the original enzyme immunoassays, antigens were coated on the plastic followed by an incubation of the serum and an incubation of labeled anti-IgM antibodies parallel to an incubation with labeled anti-IgG antibodies. This setup suffers from several problems: (i) IgG competes with the IgM in the serum for the same antigens and avidity differences may bias the results; and (ii) rheumatoid factors may adsorb nonspecifically to complexed IgG thereby resulting in false positives. Class capture assays as designed for RIA were adopted by Duermeyer and van der Veen (1978) for enzyme immunoassays and are used mainly for hepatitis viruses, rubella virus, flaviviruses, and herpesviruses. The design of these assays, discussed in Chapter 4, Section A,1, has improved IgM testing significantly since competition between IgG and IgM is eliminated. This method can be shortened by labeling the antigen (Schmitz *et al.*, 1980; van Loon *et al.*, 1981). This method may have some drawbacks, however. The IgM is extracted from the serum irrespective of whether it is an antibody and the sensitivity of the test is related directly to the ratio of specific IgM antibodies to total antibodies (Heinz *et al.*, 1981).

b. Problems That May Arise in IgM Antibody Assays

Nonspecific binding is directly related to the dilution factor (law of mass action), and seems to be higher for IgM than for IgG (Meurman, 1983). This binding involves two types of reactions: (i) nonspecific adsorption of immunoglobulin to antigen preparation, and (ii) immunological reactions of nonspecific antibodies (antinuclear, anticellular) with contaminants in the antigen preparation.

The use of crude antigens is, therefore, not always possible (Forghani and Schmidt, 1979), though for some viruses (measles, mumps, herpesviruses) excellent results have been obtained with crude extracts. Nonspecific adsorption, on the other hand, can be decreased by preincubation with "inert" proteins (BSA, gelatin).

A most important cause of false positive results is the presence of rheumatoid factor, an anti-IgG antibody of the IgM class. The rheumatoid factors react with several IgG, explaining (according to the law of mass actions) why rheumatoid factors have a steeper dose–response curve (Meurman and Ziola, 1978). Dilution is, therefore, one of the measures that can be taken to decrease this interference. Figure 4.8 shows how rheumatoid factors may interfere in class-capture assays.

Rheumatoid factor production is often activated by pregnancy making diagnosis of congenital infections more difficult (Reimer *et al.*, 1977; 1978). This also depends on the virus. Meurman (1983) detected false positive IgM antibody in routine rubella serology in one of every 125 sera, whereas Hoffmann *et al.* (1979) found about 10 times less false positives by testing tick-borne encephalitis IgM (IgG antibodies to tick-borne encephalitis are rare).

It should be noted that rheumatoid factor may play an important role in the pathogenesis of rheumatoid arthritis. Tarkowski *et al.* (1984) recently developed enzyme-linked immunospot assay for enumeration of cells secreting IgG rheumatoid factor and simultaneous quantitation of the IgG RF secreted.

Rheumatoid factor interference can be eliminated to some degree in several ways. IgG can be removed when practical, e.g., with protein A, a procedure that has been found effective (Johnson and Libby, 1980). On the other hand, rheumatoid factor may be eliminated to a greater or lesser extent with polymerized IgG [heat aggregated (Meurman and Ziola, 1978; Leinikki *et al.*, 1978), glutaraldehyde (Krishna *et al.*, 1980), or IgG-latex particles (Vejtorp, 1980)]. None of these methods is completely satisfactory and all have been shown to fail in certain cases. It is important to remember for enzyme immunoassays that antibodies labeled with an enzyme tend to aggregate leading to false positives for sera with rheumatoid factor (Yolken and Leister, 1981; Yolken *et al.*, 1983).

$F(ab')_2$ does not react with rheumatoid factor since it lacks Fc, and is a prime candidate for class-capture assays. Heterologous IgM antibody reactions may occur quite often with closely related viruses (Aaskov and Davies, 1979; Burke and Nisalak, 1982), though their titer tends to be low. This requires highly purified antigens (Katze and Crowell, 1980), and may also provide a means to perform group-specific diagnosis.

A major problem in IgM diagnosis is the lack of an IgM response (young children, local infections), e.g., for respiratory syncytial virus (Welliver *et al.*, 1980), influenza A and B virus (Goldwater *et al.*, 1982), echoviruses (Reiner and Wecker, 1981), and rotaviruses (McLean *et al.*, 1980). On the other hand, IgM responses may occur in secondary responses (Harcourt *et al.*, 1980; van Loon *et al.*, 1981) in 5–75% of the cases. Moreover, a large variation in the IgM response exists and the IgM antibody may sometimes be transient, but in other cases persist up to 4 years (Stallman *et al.*, 1974). This is often related to complex infections, pregnancy, immunosuppression, or to chronic infections (Dormeyer *et al.*, 1981).

4. Viral Gastroenteritis

a. Rotaviruses

Rotaviruses infect gastrointestinal tracts of individuals of many species including man. The clinical and serological consequences are age dependent. An infection may be asymptomatic in newborn babies, without the generation of persistent antibodies (Gust *et al.*, 1977), whereas in young children, an infection results mostly in acute diarrhea and the appearance of persisting antibodies (Flewett and Babiuk, 1984). Reinfection at a later age (symptomatic or asymptomatic) raises the antibody titer. The disease occurs in both endemic and epidemic forms and is estimated to be second in frequency only to the common cold in the United States (Cukor and Blacklow, 1984). Several viruses may cause gastroenteritis and identification and characterization of the causative agent are needed to understand its epidemiology or immunology and to manage the disease. Among the viruses causing gastroenteritis, rotavirus and Norwalk virus are medically the most important etiological agents, but enteric adenoviruses, caliciviruses, astroviruses, and coronaviruses have also been implicated, though, a host of other virus(like) particles was found in diarrheal stool specimens by electron microscopy.

The classification of rotaviruses has been confusing and was, before human rotavirus could be cultivated, based on different immunological techniques. Zissis and Lambert (1978) distinguished two human serotypes. The relationship between these two types was elucidated by a rescue technique of noncultivable rotavirus by gene assortment during mixed infection with bovine rotavirus (Greenberg *et al.*, 1981). Kalica *et al.* (1981) demonstrated that the subgroup specificity is determined by the major inner capsid protein and the serotype specificity by an outer capsid glycoprotein. The serotypes (four for human rotaviruses) can be distinguished by neutralizing antibodies, whereas the subgroups can be distinguished by, e.g., ELISA. Electropherotyping is unlikely to be useful for classification (Chanock *et al.*, 1983) since many different electropherotypes can be distinguished (32 different ones detected from 149 cases by Spencer *et al.*, 1983) and different sequences do not necessarily correlate with different migration patterns. Cukor and Blacklow (1984) suggested that electropherotyping may be of use for limited epidemics or for the study of antigenic drift (evolution of new rotavirus strains). Rotavirus antigens can be localized in the cells by the direct immunoperoxidase technique (Kurstak *et al.*, 1981) with high resolution (Fig. 10.4). Recently Grom and Bernard (1985) described an enzyme-linked cell immunoassay for detection of rotavirus antigens in which the advantages of the cell culture system for virus isolation are combined with enzyme immunodetection and spectrophotometrical reading of the test.

Cultivation of human rotavirus is not adequate for diagnostic purposes since the isolation rates are low and several passages may be needed. In contrast, electron microscopy may be most useful though limited to those institutions having highly specialized equipment and personnel. Moreover, the number of specimens that can be checked is rather limited.

Solid-phase immunoassays are most useful for rotavirus detection (Yolken, 1982). In the original enzyme immunoassay, the reactivity of the test specimens was compared to the reactivity of known negative stool samples (Yolken *et al.*, 1977). Neutralization tests with high-titered anti-rotavirus serum was thought to be unnecessary for routine testing. However, it became clear that the frequency of nonspecific reactions was high (Krause *et al.*, 1983). Different modifications have been made to improve reliability, such as adding normal serum, reducing or chelating agents, protease inhibition, or pH neutralization (Hammond *et al.*, 1982; Hogg and Davidson, 1982; Hovi *et al.*, 1982; Beards *et al.*, 1984). Krause *et al.* (1983) found the commercial Rotazyme EIA to be unreliable for neonatal specimens. The Rotazyme test seems less sensitive than indirect enzyme immunoassays (Cukor and Blacklow, 1984). The Rotazyme test from Abbott Laboratories (Table 10.3) is based on the use of anti-simian rotavirus serum. The assay depends on the fact that simian (SA11) and human rotaviruses share a common group antigen. Cukor *et al.* (1984) prepared a monoclonal antibody to the common rotavirus antigen (the sixth viral gene product). This reagent was superior in both sensitivity and specificity to the Rotazyme. Their study did not, however, include diarrheal stools of neonates. Coulson and Holmes (1984) developed an improved enzyme-linked immunosorbent assay for the detection of rotaviruses in feces of neonates.

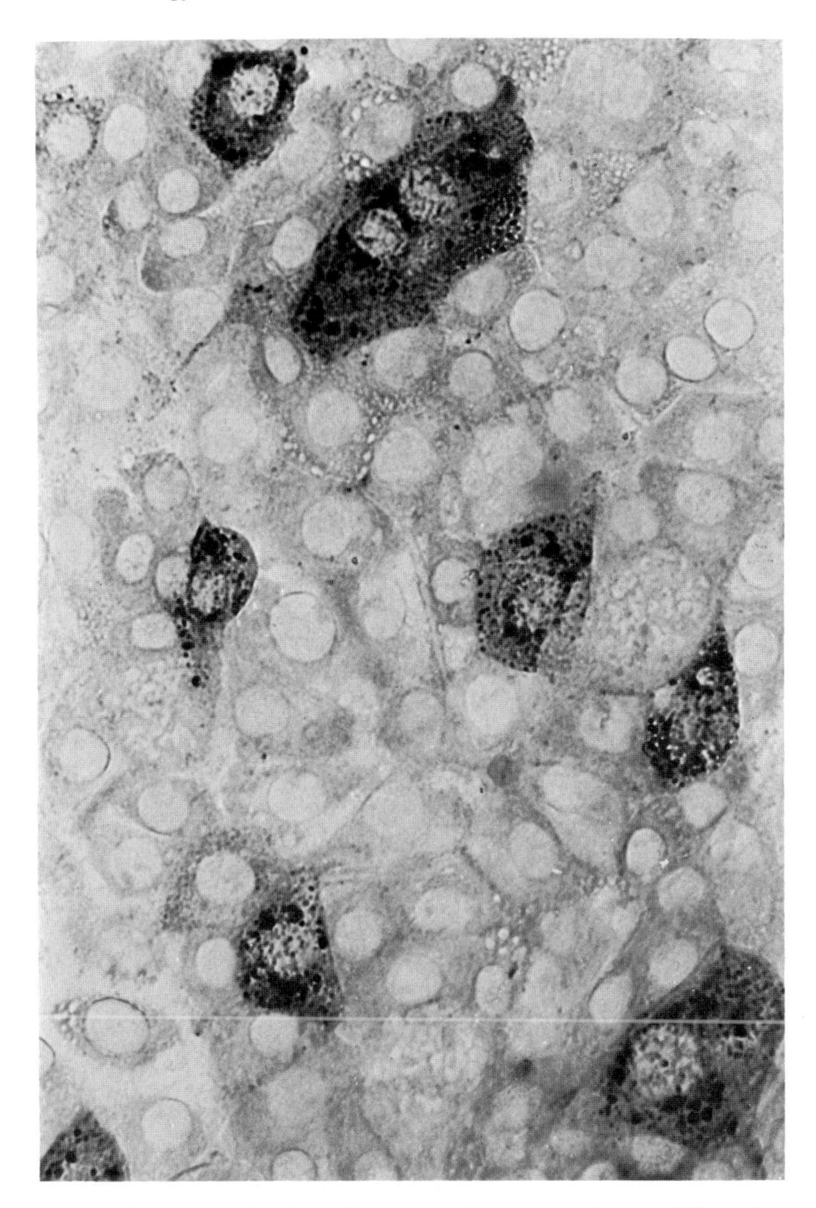

Fig. 10.4. Intracytoplasmic localization of calf rotavirus antigens in BSC-1-infected cells using direct immunoperoxidase staining.

TABLE 10.3
Detection of Rotavirus Antigen by Solid-Phase Enzyme Immunoassay[a]

Reagents
1. Anti-rotavirus (guinea pig)-coated beads
2. Anti-rotavirus (rabbit) –peroxidase (horseradish) conjugate. Minimum concentration 0.2 μg/ml in phosphate-buffered saline (PBS)
3. Inactivated simian rotavirus SA-11 in PBS as positive control
4. Diluent 0.01 *M* PBS
5. *o*-Phenylenediamne–2HCl (OPD substrate solution)
6. Diluent for OPD: citrate-phosphate buffer containing 0.02% hydrogen peroxide (H_2O_2)
7. Stopping reagent: 1 *N* sulfuric acid

Assay procedure

First incubation
1. Pipet 200 μl of each control and each diluted specimen into appropriate reaction tubes or tray wells
2. Add one bead to each tube or well containing specimen or control
3. Apply cover seal gently, tap the tube or tray to cover bead with sample and remove any trapped air bubbles
4. Incubate at 37°C for 60 min
5. Discard cover seal. Wash each bead three times with 4 to 6 ml of distilled or deionized water

Second incubation
6. Add 200 μl to enzyme conjugate into each tube or well
7. Apply new cover seal. Tap tube or tray gently
8. Incubate at 37°C for 60 min
9. Prepare OPD substrate solution (5 to 10 min prior to color development)
10. Discard cover seal. Wash each bead three times with 4 to 6 ml of distilled of deionized water. Remove all excess liquid from the tube or tray by aspiration or blotting

Color development
11. Immediately transfer beads to EIA assay tube
12. Add 300 μl OPD substrate solution to each assay tube and two empty tubes (substrate blanks)
13. Cover and incubate at room temperature (15–30°C) for 30 min
14. Immediately compare the color intensity of each tube to the color chart (visual test: do not add acid). Add 1.0 ml 1 *N* sulfuric acid to each tube. Agitate to mix. Blank spectrophotometer with a substrate blank at 492 nm. Determine absorbance of controls and specimens at 492 nm (for photometric method select a wavelength of 492 nm on a spectrophotometer or use the Quantum analyzer—Abbott).
15. Determine cutoff value by calculating the mean of absorbance of the controls (none of the individual controls should differ more than 0.5 or 1.5 time the mean). Add 0.075 to obtain the cutoff value. The gray zone is defined between 0.75 and 1.25 times the cutoff value

[a] Rotazyme II test according to Abbott Laboratories.

For this purpose the rabbit antiserum against SA11 rotavirus was incorporated as both coating and detector antibody, and rotavirus-negative rabbit serum was applied as a coating antibody control to eliminate false positive results. A pretreatment of stools with 0.25 *M* EDTA increased both specificity and sensitivity of the assay.

Bishop *et al.* (1984) expressed the activity of rotavirus antibody in human sera as units derived from a standard curve, a method successfully adapted by de Savigny and Voller (1980). They found this to be of particular value for longitudinal epidemiological studies of the effects of neonatal rotavirus infections.

Recently, a new viral agent, which is morphologically indistinguishable from rotaviruses but antigenically different (no gs antigen), has been observed (Rodger *et al.*, 1982; Nicolas *et al.*, 1983). The medical importance of these "pararotaviruses" is not yet clear and assays have not yet been devised.

i. Enzyme immunoassay of rotaviruses

The solid-phase enzyme immunoassay technique described here (Table 10.3) follows the sandwich principle (Rubenstein *et al.*, 1980) and is marketed by Abbott Laboratories. In this method, plastic beads are coated with guinea pig antibody to rotavirus. Rotavirus in the sample (a suspension of an aliquot of patient feces) added to these beads will be immobilized by the guinea pig antibody on the solid phase. After washing, antirotavirus conjugated with peroxidase is added. This antibody is bound if virus is present and will result in enzymatic activity on the addition of substrate.

Yolken *et al.* (1978) and Sheridan *et al.* (1981) found that the immobilization of guinea pig antibodies gave more reliable results than direct adsorption of rotavirus antigens. However, for the detection of antibodies to rotaviruses the adsorption of antigens to the solid phase is desirable. In an epidemiological survey Ghose *et al.* (1978) used antigens directly adsorbed to the solid phase. Inouye *et al.* (1984) found that rotavirus antigens could be efficiently adsorbed to the solid phase if pretreated with chaotropic agents (1 *M* NaSCN, 3 *M* guanidine–HCl), but that, in order to detect fecal antibodies, an excess or proteins (e.g., fetal calf serum) should be added to protect the antigens from proteolytic activity of feces. In a recent investigation Beards *et al.* (1984) observed that the inclusion of 0.01 *M* EDTA into the buffer to dilute the antigens increased the detectability of the assay considerably. This is probably due to the conversion of the rotavirus particles to an incomplete form, thus exposing the rotavirus group antigen (Cohen, 1979). EDTA may thus be effective in reducing false negative results in samples containing predominantly complete particles (Beards and Bryden, 1981). EDTA should be used at high concentrations since, particularly in neonates, samples may have high calcium ion levels.

Irradiated plastic has a higher antibody binding capacity than other plastic, though pretreatment for 1 hr with 2% glutaraldehyde raises the performance (Beards *et al.*, 1984). Brandt *et al.* (1981) reported that their ELISA had 73% false positives. Results of Beards *et al.* (1984) suggest that these false positive results depend on the reagents used, in particular rheumatoid factors.

ii. Detection of rotavirus antigens by enzyme immunoassays

The plastic is coated with a hyperimmune guinea pig rotavirus serum diluted 10,000 times in 0.1 *M* carbonate buffer, pH 9.8, and incubated overnight at 4°C or 2 hr at 37°C. The plastic can be allowed to dry and stored until use.

In preliminary tests the optimal temperature of the antigen–antibody interaction is determined. This may vary widely, e.g., some antibodies react best at room temperature but will not react at 37°C, whereas others, such as those of the commercial kit of Abbott Laboratories, react best at 45°C.

Before the assay, all reagents are brought to room temperature. For plates, an incubator is used, whereas for tubes or plastic beads in tubes a water bath is the usual choice.

In the present antibody (guinea pig)-coated plastic bead assay, 0.2 ml of controls and diluted specimens (10% v/v fecal extracts in PBS, clarified by centrifugation; immediately before use further diluted in PBS containing 0.025 *M* EDTA) is pipetted into the appropriate tubes and one coated bead is dispensed into each tube and mixed by tapping. These tubes are covered and incubated for 3 hr at 45°C. The beads are then rinsed four times with 5 ml water. The conjugate [anti-rotavirus antibodies (rabbit) labeled with peroxidase] is added (0.2 ml/bead) and the covered tube is incubated for 1 hr at 45°C. The beads are washed six times with 5 ml of water and transferred to other tubes (''assay tubes''). *o*-Phenylenediamine is added 2 mg/ml to 0.2 *M* citrate-phosphate buffer, pH 5.0, containing 0.02% hydrogen peroxide, just before use. The substrate (0.2 ml) is added to the beads and incubated for 15 min at room temperature. The optical density may be read against a color chart or, after 1 ml of 1 *N* sulfuric acid is added, at 492 nm. The detailed procedure to detect rotavirus antigen using Rotazyme kit from Abbott Laboratories is described in Table 10.3.

iii. Detection of rotavirus antibodies by enzyme immunoassays

After 6–7 days, rotavirus antigens are difficult to detect. The following method (Inouye *et al.*, 1984) can then be used.

Purified inner capsid particles (density, 1.38 g/cm^3) are mixed with an equal volume of 2 *M* NaSCN and incubated for 15 min at room temperature. The antigen is then diluted to 1 μg/ml with 0.05 *M* carbonate buffer, pH 9.6, dispensed into the wells, and incubated overnight at 4°C. After washing (three times with PBS containing 0.05% Tween 20; PBS-T), sample of feces diluted in PBS-T (containing 1% BSA or 10% fetal calf serum; FCS) is added and incubated for 2 hr at room temperature. After washing the wells, conjugate in PBS-T (1% BSA) is added and incubated for 1 hr at room temperature. In order to detect fecal antibodies, an excess of proteins (e.g., fetal calf serum) is added to protect rotavirus antigens from proteolytic activity of feces.

The enzyme is revealed as described above. BSA was much less effective than FCS in preventing proteolysis by constituents of the feces. Though antibodies from a different species could be used to trap the rotavirus antigens on the solid phase, cross-reactivity with the sample antibodies may be a problem.

The detection of rotavirus-specific IgG antibodies by an indirect immunoperoxidase method and enzyme-linked immunosorbent assay is also proposed using as antigen, SA11 rotavirus infected MA 104 cells. In parallel, rotavirus-specific IgG antibodies were determined by ELISA (Zentner *et al.*, 1985).

b. Norwalk Virus

About 10 years ago the Norwalk virus was visualized by immunoelectron microscopy in stool specimens obtained from an outbreak of gastroenteritis (Kapikian *et al.*, 1972). This virus has the same physical properties and morphology as the parvoviruses. However, if the observations of Greenberg *et al.* (1981) that this virus contains only one protein with an M_r of 66,000 are confirmed, it is more probably a member of the caliciviruses (RNA containing).

The virus is shed in low titers (Greenberg *et al.*, 1981), making it difficult to prepare hyperimmune animal serum reagents for diagnosis.

Currently, the RIA test procedures used rely on fecal and serum samples from experimentally infected volunteers.

c. *Other Gastroenteritis Causing Viruses (Adenoviruses, Astroviruses, Caliciviruses, Coronaviruses)*

New serotypes of adenoviruses have been isolated from stool specimens from cases of infantile gastroenteritis (de Jong *et al.*, 1983) and are recognized to be, after rotavirus, the most common agent of infantile gastroenteritis (Brandt *et al.*, 1983). Though these viruses do not grow in standard tissue cultures, they can be grown in Graham 293 cells (Takiff *et al.*, 1981), and production of reagents for enzyme immunoassays is now possible.

About 90% of adults have antibodies to calicivirus (Sakuma *et al.*, 1981). The detection of these viruses has relied on electron microscopy, but recently a solid-phase RIA was developed (Nakata *et al.*, 1983) and enzyme immunoassays can be expected soon.

Coronavirus-like particles have also been associated with gastroenteritis, though its definite role cannot be assessed since the particles were observed in similar proportion in healthy individuals (Cukor and Blacklow, 1984).

Astroviruses cannot be grown *in vitro*, making it difficult to prepare diagnostic reagents, so diagnosis still relies primarily on immunoelectron microscopy (Kanno *et al.*, 1982).

5. Respiratory Tract Infections

Several viruses are etiologically important in association with these syndromes, though as with other clinical symptoms many different viruses may induce such infections. However, to prevent a reiteration of many different possible viruses as a cause for each syndrome only the most important will be discussed.

Upper respiratory tract infections by viruses may lead to common cold (undifferentiated), febrile upper respiratory illness, pharyngitis, and pharyngoconjunctival fever. The lower respiratory tract infections can lead to laryngotracheobronchitis or bronchiolitis (both infants and children), bronchitis (children and adults), and pneumonia. The viruses which may be involved are listed in Table 10.4, though this list is probably incomplete.

The majority of upper respiratory tract infections are caused by viruses and occur in individuals of all ages. Croup or acute laryngotracheitis is

TABLE 10.4
Viruses Involved in Respiratory Infections[a]

Syndrome	Common agent	Less common agent	Sometimes
Common cold	Rhinoviruses	Coronaviruses Parainfluenza virus Coxsackievirus (A21) Echovirus	Reoviruses (1–3)
Febrile upper respiratory illness	Coronavirus Influenza virus A, B	Adenoviruses (1–5, 7, 14, 21) Parainfluenza viruses	Echoviruses Reoviruses (1–3)
Pharyngitis	Adenovirus (1, 3, 4, 5, 7)	Coxsackieviruses (A2, 4, 5, 6, 8, 10`	Herpesvirus
Pharyngoconjuctival fever	Adenovirus (3, 7, 14)		
Influenza	Influenza A, B		
Laryngotracheo-bronchitis	Parainfluenza 1, 2, 3	Influenza A, B	Respiratory syncytial virus
Bronchiolitis and bronchitis	Respirtory syncytial virus Influenza viruses A, B	Adenoviruses (1, 2, 3, 5, 7)	
Pneumonia	Influenza A, B Respiratory syncytial virus Parainfluenza viruses (1–3)	Adenoviruses (3, 4, 7, 7a)	Cocksackieviruses Echoviruses Rubella virus Herpesviruses (Varicellae)

[a] Adapted from Mufson (1978).

primarily associated with parainfluenza 1. Among the lower respiratory tract infections, bacterial infections are much more important (perhaps one-third of the total), particularly in the case of pneumonia. Viral pneumonia occurs primarily in the very young, the old, and in those with respiratory handicaps. In addition to these viruses, measles, chickenpox, and rubella start as respiratory tract infections.

a. Influenza Viruses

Resistance to infection with influenza A is associated with antibodies to hemagglutinin and neuramidase (Virelizor *et al.*, 1979) but not with antibodies to the internal protein. The traditional hemagglutinin and neuramidase inhibition tests are relatively insensitive. Turner *et al.* (1982) compared the ELISA to hemagglutinin inhibition (AHI) tests and found increases in geometric mean antibody titers with ELISA to be 16 to 71-fold, whereas for HAI this was only 3- to 10-fold (depending on antigen used for assay) on infection. Julkunen *et al.* (1985) found also that the efficacy of enzyme immunoassay in detecting diagnostic antibody uses to influenza A and B viruses was higher when compared with complement fixation and hemagglutination inhibition tests. The presence of antibodies can be tested in sera or in nasal washings. Murphy *et al.* (1981) detected with ELISA primarily IgG-isotype antibodies to influenza in sera and IgA-isotype antibodies in nasal washings. The antibodies manifested a hemagglutin subgroup specificity. Therefore, Murphy *et al.* (1981) devised a sensitive ELISA for hemagglutinin instead of whole virus. Similarly, Khan *et al.* (1982a) developed an ELISA for influenza virus neuramidase-specific antibodies. Khan *et al.* (1982b) also developed an ELISA to M protein (isolated by SDS–gel chromatography), antibodies to which are generally present in patients with severe illness. Hammond *et al.* (1980) noted a lack of subtype specificity by ELISA. They used end-point titers to find differences, but as pointed out by Madore *et al.* (1983), this method may be inadequate. Madore *et al.* (1983) observed significantly greater antibody concentrations with influenza infection when the areas under the dose-response curves were considered (or the area generated between the two response curves) than with the end-point method. Many of the other data processing methods also rely on the assumption that the dose–response curves are parallel, but in fact the curve may become steeper without an increase in the end-point.

In addition to influenza A and B, influenza C has been recognized as a human pathogen, but information has been limited because of difficulties in propagating the virus. Austin *et al.* (1978) succeeded in adapting the virus to grow in the allantoic cavity of embryonated eggs. Large quantities of reagents can thus be prepared, and an ELISA has been designed for this virus (Troisi and Monto, 1981). The test appeared to be more sensitive after removal of potentially interfering chicken egg albumin antibodies. Heat treatment (56°C for 30 min) of the serum in some cases was found to increase readings. The virus seems to be ubiquitous and apparently infections occur frequently.

The differentiation of strains of influenza viruses requires the use of their constituent proteins instead of whole virus (Al-Kaissi and Mostratos, 1982). Influenza viruses adsorbed to the surface of polystyrene were found ruptured by the adsorption process and the exposed nucleoprotein was able to react with antibody. This is probably why the specificity of enzyme immunoassays resembles that of the complement fixation test rather than the hemagglutination test (Hammond *et al.*, 1980). However, hemagglutinin or neuramidase preparations are not generally available. Adsorption of complete virus to polystyrene inactivated the neuramidase of at least one strain (Al-Kaissi and Mostratos, 1982).

Most respiratory diseases are diagnosed by isolating the agent in cell cultures or embryonated eggs. Detection of antigens in cells by immunofluorescence or immunoperoxidase can speed up diagnosis but is not as sensitive. Harmon and Pawlik (1982) developed an ELISA for the direct detection of influenza A antigens without prior cultivation. Virus-specific guinea pig immunoglobulins were coated to polystyrene microtiter plates and the antigens were captured from the specimens. This was followed by successive incubations with virus-specific rabbit antiserum and goat anti-rabbit immunoglobulin antibodies conjugated with peroxidase. The detection limit of the standard test equalled about $10^{3.5}$ $TCID_{50}$ units. The inclusion of 1% gelatin in the last two incubations increased sensitivity 3- to 6-fold not only because of decreased background but also, interestingly, because of increased readings. This method detected 50–60% of the isolation-positive clinical specimens. Harmon and Pawlik (1982) obtained similar results for the detection of respiratory adenovirus. It is important to note that some patients with clinical findings of influenza virus infection do not have virus shedding (Ksiazek *et al.*, 1980). It should be possible to increase the sensitivity of the antigen detection by ELISA significantly with some of the new techniques available. Though hemagglutin preparations are not generally available it is relatively simple to purify them (Phelan *et al.*, 1980).

b. Parainfluenza Viruses

Parainfluenza viruses are a common cause of acute respiratory illness in infants and children. The traditional laboratory diagnosis involves the isolation of the virus from throat swabs or nasopharyngeal secretions and cultivation in primary kidney cell cultures. However, the difficulties in transportation and storage of parainfluenza viruses are notorious; infectivity is very rapidly lost. The identification of the virus produced in the cells is achieved by hemadsorption of guinea pig erythrocytes or hemagglutinin inhibition. These methods are expensive and time consuming. On the other hand, the immune response may also take 2–3 weeks. Wong *et al.* (1982) found an immunofluorescence assay for the antigen in nasopharyngeal epithelial cells to be more reliable and faster.

Sarkkinen *et al.* (1981a,b) compared immunofluorescence, radioimmunoassay, and enzyme immunoassay for the detection of parainfluenza type 2 and found excellent agreement. They noted that nasopharyngeal washings contained too little antigen in about 50% of the positive cases, both with radio- and enzyme immunoassays. In contrast, aspirating the secretions with a mucous extractor through the nostrils from the nasopharynx yielded specimens which gave, after sonication for 3 min with a sonicator with a microtip, excellent results. No false positives were observed, though the authors advised blocking tests for low absorbance values.

Ukkonen *et al.* (1980) devised an ELISA for parainfluenza-type 1 antibodies. They did not detect an increase in IgM antibodies in cases of increasing antibody titers with the complement fixation method. Van der Logt *et al.* (1982) confirmed this observation, detecting IgM antibody in a low proportion of the infections. Recently, Julkunen (1984) compared enzyme immunoassay with the complement fixation test for parainfluenza virus antibodies. In this study performed on 180 patients with respiratory symptoms, EIA was found to bring significantly more diagnostic rises in parainfluenza virus type 1, 2, and 3 antibodies compared to the CF test. The enzyme immunoassay could be the diagnostic procedure of choice in parainfluenza virus infections, where reinfection are frequent, with the virus replication usually limited to the upper respiratory tract, and consequently, the antibody responses often remaining faint.

The fact that for enzyme immunoassays no viable virus is required is a great advantage for parainfluenza viruses since these are inactivated so rapidly. This was stressed again by Parkinson *et al.* (1982) who could detect parainfluenza antigens in all culture residuals from which infectious virus could no longer be recovered. The sensitivity of this indirect enzyme immunoassay equalled about 0.25 hemagglutinin units for parainfluenza virus type 1, whereas for types 2 and 3 the sensitivity of the enzyme immunoassay equalled about 0.01 hemagglutinin units.

c. *Respiratory Syncytial Virus (RSV)*

RSV, which infects primarily the young, is diagnosed in the laboratory by rapid inoculation of cell cultures. This is possible only in laboratories near or in hospitals and has the disadvantage that confirmation takes a few days. Enzyme immunoassays are, therefore, prime candidates for diagnostic purposes and have been widely applied (Chao *et al.*, 1979; Sarkkinen *et al.*, 1981; Hornsleth *et al.*, 1981, 1982; McIntosh *et al.*, 1982; and Meurman *et al.*, 1984a,b).

McIntosh *et al.* (1982) observed that specimen transportation is not a great problem for enzyme immunoassays since specimens left for days at room temperature will be equally reactive. Chao *et al.* (1979), Hornsleth *et al.* (1982), and Meurman *et al.* (1984a) compared the immunoassays for the antigen with virus isolation procedures from cell cultures and showed that the clinical sensitivity of these assays is about 80%. Hornsleth *et al.* (1982) observed, however, that enzyme immunoassays is more sensitive for children under 6 months of age, an observation confirmed by Meurman *et al.* (1984a).

Serological examination with enzyme immunoassays should be reevaluated if clinical symptoms suggest an RSV-induced disease, since false negatives occur frequently (Meurman *et al.*, 1984b). IgG levels are relatively high with significant increases over 2-week periods and testing with enzyme immunoassays is clearly superior to complement fixation (Hornsleth *et al.*, 1984). In very young patients, maternal antibodies may interfere. IgM antibodies which are an indication of acute infections of many viruses are often absent in RSV infections, though the ability to produce these IgM antibodies seems to increase with age (Welliver *et al.*, 1980; Meurman *et al.*, 1984b).

IgA antibodies are frequently absent and their presence in a single test is not indicative of a recent infection since the antibodies may persist over prolonged periods (Welliver *et al.*, 1980).

d. Coronaviruses

Kraaijeveld *et al.* (1980) described an ELISA to detect antibodies to human coronavirus. Macnaughton *et al.* (1983) developed an assay to detect coronavirus antigens in nose swabs, throat swabs, and nasopharyngeal aspirates from children with naturally acquired respiratory infections. Samples with at least 10^5 particles/ml can be detected. About twice as many positives were obtained with nose-swab specimens then with throat swabs or nasopharyngeal aspirates.

Schmidt and Kenny (1982) have shown that human coronaviruses contain three different antigens. Schmidt (1984) developed an ELISA to investigate the nature of these polypeptides. No cross-reactivity was found between the antigens of the coronaviruses 229E and OC43. Most prevalent human antibodies were found directed against virion peplomers.

e. Measles

Measles starts as respiratory viral infections and has long been recognized as one of the major diseases of childhood (Marusyk and Tyrrell,

1984), often involving the central nervous system (Norrby, 1978). Complications, especially in the malnourished child, include pneumonia, otitis media, diarrhea, encephalitis, and subacute sclerosing panencephalitis (SSPE).

An enzyme immunoassay test (Enzygnost) was developed by Hoechst Institute which is based on the detection of specific IgM antibodies. In measles infection IgM antibodies appear within the first days after the onset of rash and persist about 50–90 days. Prolonged IgM antibody responses are also observed in a minority of patients with SSPE. In postmeasles encephalitis prolonged IgG antibody response may occur. SSPE afflicted patients always show a 10 to 100 times higher level of specific IgG then do convalescents from acute measles infection.

In Hoechst Enzygnost test, antibodies to viral antigen (IgM, IgG) present in the serum sample are incubated with the respective antigen fixed in the wells of microplates. Antibodies specific to the virus antigen are bound to the solid-phase antigen after incubation for 1 hr at 37°C. Consequently anti-human IgG (from rabbit, γ chain specific)/or anti-human IgM (from rabbit, μ chain specific) conjugated with alkaline phosphatase is incubated with the previously formed solid-phase antigen–antibody complex.

The anti-human IgG conjugate or anti-human IgM conjugate is bound to the antigen–antibody complex forming a sandwich. The addition of chromogenic substrate (*p*-nitrophenyl phosphate) dissolved in substrate buffer solution results in a yellowish-green color if specific antibodies are present in the patient's sample. The enzymatic reaction is stopped by addition of 0.05 ml of 2 *N* NaOH per well. Photometric evaluation of results obtained, including reading of the plate and calculating and printing of results, is automatically performed by a Behring ELISA-Processor M.

6. Prenatal Infections

The most common cause of prenatal infections in man are the cytomegalo- and rubella viruses. The contamination rate for cytomegalovirus ranges from 60% of the adult population in developed countries to 100% in developing countries or in low socioeconomic groups (Krech and Jung, 1971). Like other herpesviruses, cytomegalovirus can persist latently. The wide variety of (often subclinical) symptoms makes diagnosis difficult. The main effects of intrauterine infection are hepatosplenomegaly and diseases of the central nervous system. Rubella is usually benign but problems of infection may arise during the first trimester of pregnancy. About 80–90% of women of childbearing age have antibody protection.

a. *Cytomegalovirus (CMV)*

IgM antibodies to CMV are excellent markers for congenital or recently acquired CMV infections. Immunoperoxidase procedures (Gerna and Chambers, 1977a,b) and enzyme immunoassays (Schmitz *et al.*, 1977), particularly class-capture assay, (Schmitz *et al.*, 1980; van Loon *et al.*, 1981) have been used to explain this EIA phenomenon. However, IgM antibody production can be highly variable following secondary infections (Sutherland and Briggs, 1983). Removal of rheumatoid factor is essential to prevent false positives (Kangro *et al.*, 1984).

Booth *et al.* (1979, 1982) developed a CMV IgG-antibody test and compared it to other serological methods. Poor correlation was found between RIA and EIA titers. It seems that different antibodies are labeled for these two tests. The anticomplement immunoperoxidase method offers a sensitive method for the specific detection of intranuclear inclusion bodies in infected cells (Fig. 10.5).

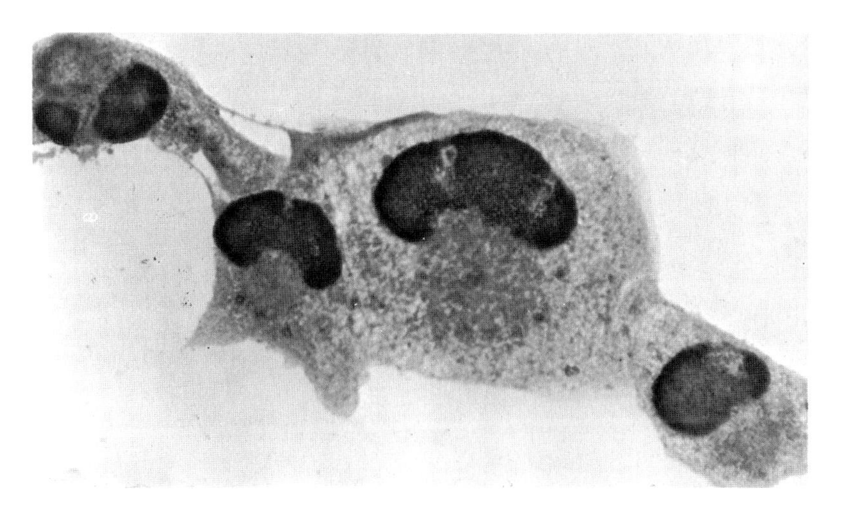

Fig. 10.5. Anticomplement immunoperoxidase staining of intranuclear inclusion bodies after infection of human embryonic lung cells with the AD169 strain of cytomegalovirus (CMV). CMV induces the synthesis of Fc receptors which react nonspecifically with CMV-negative sera in the indirect immunoperoxidase method. These false positive results are avoided with the anticomplement method confirming results with the EBNA detection (Fig. 4.8). The intranuclear inclusion bodies are specifically stained, whereas the intracytoplasmic inclusion body can be observed by the insignificant background staining. The positive CMV serum was diluted 40× and 1:10 fresh CMV-negative serum was added for complement. The conjugate was diluted 100×. Courtesy of Dr. Yoichi Minamishima and Dr. Seigo Yamamoto with permission to reproduce.

Kurstak *et al.* (1971) used light and electron microscope immunoperoxidase staining for the detection of cytomegalovirus antigens in infected human fibroblasts. It was possible to shorten substantially the time necessary to diagnose infection in comparison with the CPE method. Recently, Sarov and Haikin (1983) proposed a potential application of the indirect immunoperoxidase IgA assay for serodiagnosis of CMV infection using sequential serum samples of patients. Since the persistence of IgA antibodies depends on the virus involved, individual variations among the patients, and, most importantly, on the sensitivity of the method used to detect specific IgA, the immunoperoxidase IgA assay was suggested. In another work Singer *et al.* (1985) adopted the immunoperoxidase method for detection of serum IgM antibodies to CMV. The antigen consisted of CMV infected human embryonic fibroblasts or isolated nuclei. Rabbit anti-human IgM peroxidase conjugate was used to detect IgM bound to viral antigen. The indirect immunoperoxidase was also recommended as an advantageous method in both diagnosing and evaluating CMV mononucleosis (Abramowitz *et al.*, 1982).

b. Rubella Virus

Class-capture assay for rubella-specific IgM antibody proved a simple, sensitive, and specific alternative for IgM anti-rubella testing and compared favorably to other approaches (Isaac and Payne, 1982). This method, developed by Duermeyer and van der Veen (1978), avoids the time consuming or expensive physical separation of IgG from IgM by sucrose density gradient centrifugation or exclusion chromatography (Vesikari and Vaheri, 1968; Gupta *et al.*, 1971). Class-capture assays using $F(ab')_2$ conjugates proved reliable and sensitive (Isaac and Payne, 1982). Removal of nonspecific inhibitors is not necessary and volumes of only 5 μl of sample are required. The clinical validation of antibody-capture anti-rubella IgM-immunoassay was performed by Wielaard *et al.* (1985). In their study using monoclonal antibodies IgM was found by EIA in 99 sera on total of 103 sera, known to contain anti-rubella IgM by a sucrose density gradient method. In a second study, 16 out of 17 acute rubella infections were detected by the IgM-EIA. The specificity of the antibody-capture EIA was high and no interference was seen in 60 rheumatoid-factor positive sera, in 100 highly positive IgG sera. Recently Bonfanti *et al.* (1985) developed a new solid phase enzyme immunoassay for detection of rubella-specific IgG. This procedure uses polystyrene microtiter strips coated with rabbit anti-human IgG immunoglobulins as the solid phase and an enzyme-labeled semipurified rubella antigen as indicator. An advantage of their direct enzyme immunoassay is that the same method and some labeled antigen can be used to test for different classes of immunoglobulins using simply solid phase coated with capture antibodies of different chain specificity.

Best *et al.* (1984) demonstrated problems that may occur when detecting rubella-specific IgM with the Rubazyme-M (Abbott Diagnostics) and the M-antibody capture radioimmunoassay (MACRIA). They recommended that results for rubella-specific IgM should be interpreted with caution. Rubazyme-M is not (currently) recommended by the manufacturer for the diagnosis of congenital rubella by examination of cord blood. This was supported by Best *et al.* (1984) who found 2/20 false negatives from neonates with evidence of intrauterine infection. In contrast, false positive results due to rheumatoid factors or anti-nuclear antibody were not obtained. Rubella-specific IgM detected in sera from patients with infectious mononucleosis may be due to EBV stimulation of B lymphocytes specific for rubella or result from an incorporation of cellular antigens in the lipoprotein envelope of the rubella virus used as antigen (Morgan-Capner *et al.*, 1983).

Kurtz and Malic (1981) showed that a disadvantage of the ELISA system is that the presence of rheumatoid factor may cause false positive results. The RF can attach the patient's IgG rubella antibody to the solid phase, or it can react with altered IgG in the conjugate. They overcome the latter problem by removing altered IgG from the conjugate or by conjugating only the Fab_2 portion. Another possibility to assure the specificity is to perform a duplicate test on each serum, omitting the addition of antigen.

The diagnosis of rubella virus in clinical materials with the use of the immunoperoxidase method was also satisfactory when patients' specimens were inoculated with BHK-21 hamster kidley cells (Schmidt *et al.*, 1981).

7. Herpetic Infections

a. *Herpes Simplex Virus (HSV)*

Numerous techniques have been developed for the identification and typing of herpes simplex virus. The serological techniques include immunofluorescence (Nahmias *et al.*, 1971), immunoperoxidase (Fig. 10.6) (Benjamin, 1974; Kurstak and Kurstak, 1974; Morisset *et al.*, 1974; Gerna and Chambers, 1977a), hemagglutination (Bernstein and Stewart, 1971), mixed agglutination (Ito and Bairon, 1974), microneutralization, counterimmunoelectrophoresis (Jeansson, 1974), radioimmunoassay (Forghani

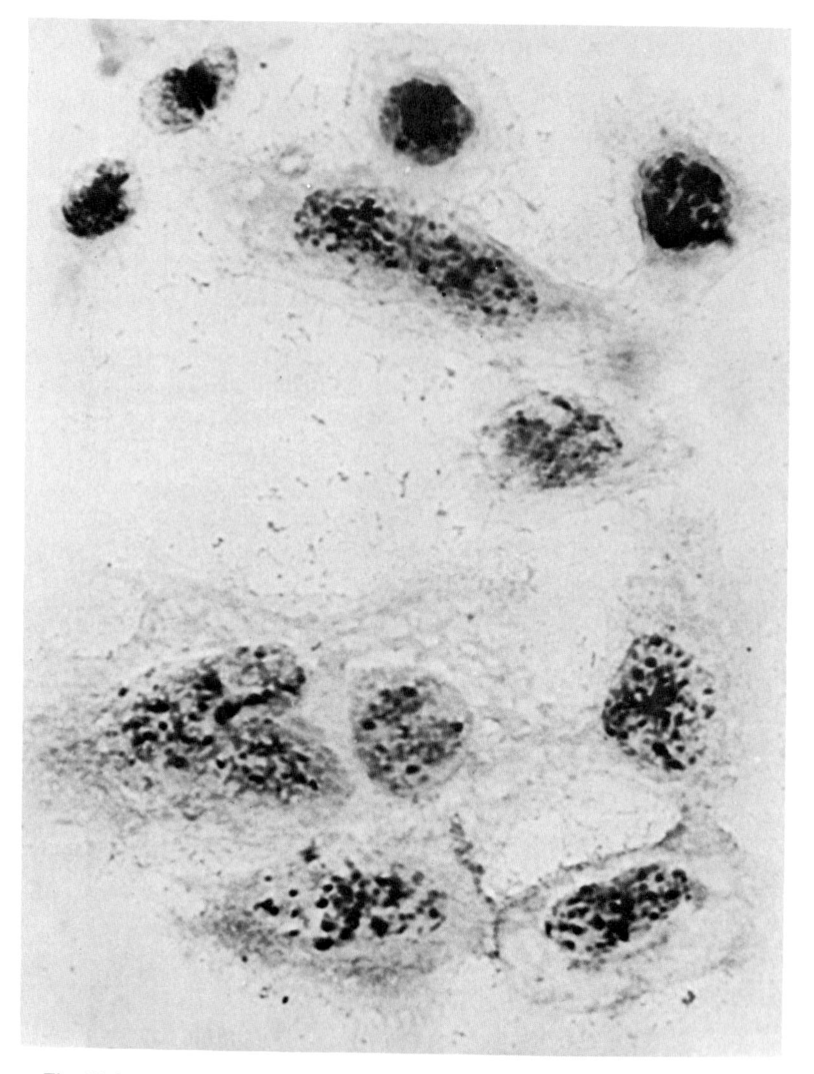

Fig. 10.6. Herpes simplex virus type 2 antigens are well localized in human cells Hep-2 by the direct immunoperoxidase method.

et al., 1974), and enzyme immunoassay (Vestergaard and Jensen, 1980; Frame *et al.*, 1984). Contrary to initial beliefs, HSV type 1 is a frequent cause of genital lesions (38% of isolates; Balachandran *et al.*, 1982). This is important since HSV type 1 infections differ from HSV type 2 infections with respect to recurrence and response to chemotherapy and their association with cervical carcinoma (De Clercq *et al.*, 1980; Rawls *et al.*, 1980; Reeves *et al.*, 1981).

Frame *et al.* (1984) developed a double-antibody enzyme immunoassay for HSV typing by employing a polyclonal rabbit capture antiserum with type-common and type 2-specific monoclonal antibodies (Killington *et al.*, 1981) as detectors. This assay is 100% sensitive and 100% specific compared with cell culture and restriction endonuclease analysis. A typing index of 1.5 (ratio of optical density with the HSV-type common antibody to the optical density with the type 2-specific antibody) had been selected for distinguishing type 2 isolates from type 1. HSV-2 antigens can also be detected with the direct immunoperoxidase method on cytosmears of cervical scrapings (Fig. 10.7), as well as HSV-1 antigens in the bullae of typical erythema multiform lesions after mild, recurrent, localized skin infections (Major *et al.*, 1978).

b. *Varicella-Zoster Virus*

It is important to test for varicella-zoster virus which can be fatal in immunocompromised patients on cytotoxic therapy (Whreghitt *et al.*, 1984).

Forghani *et al.* (1978) developed an indirect varicella-zoster virus en-

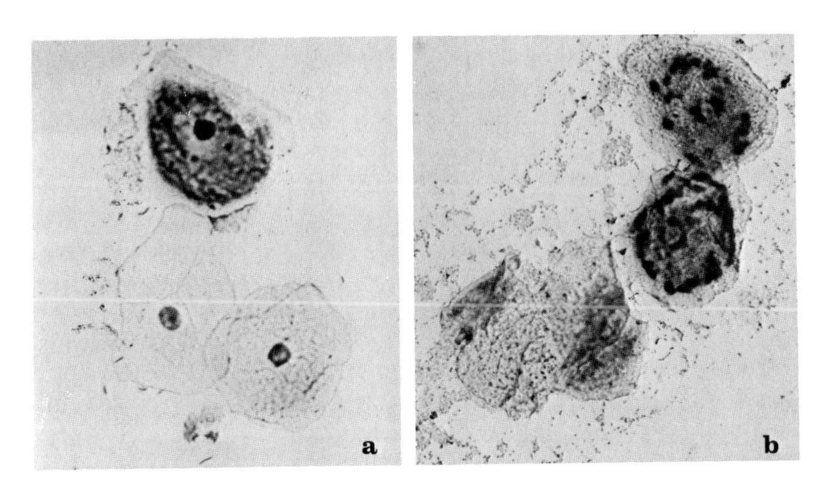

Fig. 10.7. Localization of herpes simplex virus type 2 antigens in cytosmears of cervical scraping from patients with clinical symptoms of genital infection of herpes type. Direct immunoperoxidase staining (a, b).

zyme immunoassay whereas Whreghitt *et al.* (1984) devised a competitive varicella-zoster virus enzyme immunoassay. Though the complement fixation test is usually satisfactory, it is surprisingly inadequate for the detection of anti-varicella-zoster virus antibodies in sera from hemodialysis patients. The competitive enzyme immunoassay is more specific than the indirect assay, though its potential sensitivity is lower (Kurstak, 1985). It is, however, 20 times as sensitive as complement fixation (Whreghitt *et al.*, 1984).

Cox *et al.* (1984) developed an *in situ* enzyme immunoassay to viral membrane antigen to enable the specific estimation of antibodies to varicella-zoster virus. Such a rapid screening test for these antibodies for possible inclusion in zoster immune globulin pools used to alleviate severe clinical effects of infection, particularly due to the shortage of this immune globulin, is essential. For this purpose cells were fixed with 0.05% glutaraldehyde. The *in situ* enzyme immunoassay can then be expected to be specific for membrane varicella-zoster virus antigen, whereas cross-reacting internal antigen (not expressed at host membrane) does not interfere. Also for diagnostic purpose the immunoperoxidase method was applied successfully to detect antigenic material of herpes zoster virus on cutaneous lesions of patients with dendritic keratitis (Laflamme *et al.*, 1976).

c. Epstein–Barr Virus (EBV)

The immunoperoxidase method is a sensitive and specific method for the detection of antigens related to EBV (Kurstak *et al.*, 1978; Van der Hurk and Kurstak, 1980). The anticomplement immunoperoxidase method is capable of detecting the EBV nuclear antigen (EBNA) at high serum dilution (Fig. 10.8). Similar staining patterns were obtained with the anticomplement immunofluorescence technique (Reedman and Klein, 1973). However, unlike the immunofluorescence technique, the immunoperoxidase technique allowed the detection of EBNA by the indirect method. Viral capsid antigens (VCA) and early antigen (EA) can also be detected with these sensitive methods.

Recently an enzyme-linked immunosorbent assay has been developed for the detection of antibodies to EB virus membrane antigen (MA) glycoprotein (gp340) in tamarins (Randle and Epstein, 1984). This method was found to be a thousand-fold more sensitive than conventional indirect immunofluorescent tests. A human serum, typed seronegative by the immunofluorescence technique, gave a titer of 1:400 with ELISA developed for the measurement of specific antibody to the gp340 glycoprotein subunit of Epstein–Barr virus membrane antigen. This assay is suitable to monitor

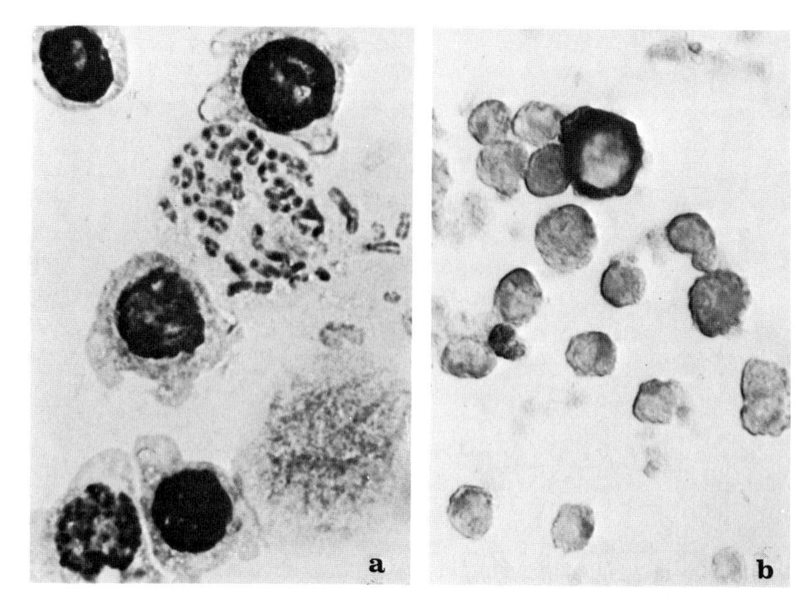

Fig. 10.8. Epstein–Barr nuclear antigen (EBNA) detection by the anticomplement immunoperoxidase (ACIP) method in Raji cells. EBNA is demonstrated in the nucleus of lymphocytes transformed by the EB virus and on the chromosomes in the cells during metaphase (a). EB virus capsid antigen (VCA) is detected in P3HR-1 cells by indirect immunoperoxidase (b).

with high sensitivity the sequential production of specific antibodies to gp340 by tamarins and for the assessment of vaccination procedures *in vivo*.

8. Papillomaviruses

Papillomaviruses (etiologic agents of warts) cannot be propagated *in vitro*. Distinct virus types have been characterized (Gissmann and zur Hausen, 1980; Gissmann *et al.*, 1982). Recently, there has been a renewed interest in this group of viruses which were previously considered to be benign. A suspicion has grown that papillomaviruses are associated with human cancer (zur Hausen, 1977; Meisels and Morin, 1981).

Braun *et al.* (1983) compared histologically diagnosed warts and cutaneous and broadly cross-reactive antiserum dysplasias by means of the immunoperoxidase technique for the frequency of occurrence and pattern of distribution of papillomavirus capsid antigen.

9. Hepatitis Viruses

Viral hepatitis can be caused by different viruses (hepatitis A, hepatitis B, hepatitis non-A, non-B, and sometimes enteroviruses, herpesviruses, cytomegaloviruses, yellow fever virus, rubella and Epstein–Barr viruses) and is a systemic disease primarily involving the liver. ''Infectious hepatitis'' (''short-incubation hepatitis'') or epidemic jaundice is generally caused by the type A virus (hepatitis A virus), whereas serum hepatitis (''long-incubation hepatitis'') is caused by the type B virus (hepatitis B virus). The major form of posttransfusion hepatitis in many developed countries is non-A, non-B hepatitis since the frequency of hepatitis B infections has been reduced due to careful screening of donor blood (Dienstag *et al.*, 1981).

a. Hepatitis A Virus Diagnosis

A still widely used method to detect hepatitis A particles is immunoelectron microscopy (Feinstone *et al.*, 1973) which allows the detection of virus from a solution containing 10^5–10^6 particles/ml. Direct detection of antigen has more value for the research laboratory than for clinical diagnosis since maximum shedding of the virus occurs before the development of liver abnormalities (Rakela and Mosley, 1977). The titers of anti-hepatitis A virus-positive antisera is generally high (Skinhøj *et al.*, 1977).

The most reliable method would detect IgM antibodies to HAV. However, it could be used only 4–12 weeks postinfection since detection usually occurs after 6 weeks (depending on the detectability of the assay). The class-capture assay (Duermeyer and van der Veen, 1978; Duermeyer *et al.*, 1979), available from Abbott Laboratories, is generally employed. Anti-IgM antibodies (anti-μ chain) are coated on the plastic. This sensitized solid phase will retrieve IgM from the serum. If among these, IgM antibodies are present, they will bind subsequently added HAV antigen. This antigen is then detected with human anti-HAV antibodies labeled with peroxidase.

Stool samples are prepared (Hollinger and Dienstag, 1980) by diluting a 20–50 g sample five times with 10 m*M* Tris–HCl buffer, pH 8.0, and homogenizing it by shaking vigorously with glass beads. The sample is then cleared by centrifugation (45 min at 1000 *g;* 4°C) and subsequently extracted with chloroform. It is then subjected to differential centrifugation for 1 hr at 10,000 rpm and 16 hr at 20,000 rpm) in a Beckman type 21 rotor (4°C). The pellet obtained is resuspended in the Tris–HCl buffer (50 ml) and layered on a cushion (10 ml of CsCl, 1.6 g/ml) and centrifuged in an

SW 27 rotor (18 hr; 25,000 rpm, 4°C). The gradient is fractionated and tested for the presence of virus. Fractions containing the virus are dialyzed against the Tris–HCl buffer and centrifuged on a 10–30% sucrose gradient (SW 27 rotor; 25,000 rpm, 4°C). The virus-containing fraction is further purified on Sepharose 2B in PBS.

Anti-HAV positive sera are acquired from infected subjects (titers frequently over 10^5) even years after infection (Skinhøj *et al.*, 1977).

b. Hepatitis B Virus Diagnosis

Hepatitis B virus yields on infection morphologically different particles, i.e., a 42-nm double-shelled sperical virus particle (originally named Dane particle), and tubular and small 20-nm-diameter particles (Peterson, 1981; Deinhardt and Gust, 1982). The surface antigen, HBsAg, is found not only in the outer shell of the Dane particle but also in the smaller and filamentous particles and consists of four polypeptides and four glycoproteins (Skelly *et al.*, 1978). The small particles contain the antigen but not the nucleic acid (Dreesman *et al.*, 1975). Analysis of the HBsAg revealed that there are two sets of mutually exclusive types (d/y and w/r) in addition to specificities common to all types (e.g., adw, adr, ayw, or ayr which can be subdivided further). These subtypes have a particular geographic distribution, e.g., adr is most common in Southeast Asia. Furthermore, HBsAg particles retain immunogenicity and antigenicity in conditions in which the infectious virus is rapidly inactivated (Zuckerman, 1981; Deinhardt and Gust, 1982).

Rapid diagnosis

Serum can be stored refrigerated or frozen. The simplest method of detection is still agar gel diffusion or rheophoresis (Jambazian and Holper, 1972). Counterimmunoelectrophoresis is rapid (Dreesman *et al.*, 1975) and more sensitive than agar gel diffusion. The reversed passive latex agglutination test is most rapid and simple for HBsAg (Kachani and Gocke, 1973) and hemagglutination (Vyas and Shulman, 1970), but false positives are common with these methods. Techniques with greatly improved sensitivity and specificity are the radioimmunoassays (Lander *et al.*, 1971; Purcell *et al.*, 1973) and enzyme immunoassays (van Weemen and Schuurs, 1971; Wei *et al.*, 1977; Kurstak *et al.*, 1984a,b).

The technique described here is based on the sandwich principle and has been adapted to detect HBsAg (subtypes ad and ay), the AUSAB[R] EIA, by Abbott Laboratories for commercial kits. Polystyrene is coated with purified HBsAg. The serum to be tested is brought in contact with the immobilized antigen which extracts the antibody. Unbound material is washed and biotinylated HBsAg is added, which binds in turn to the antibody. Avidin–peroxidase complexes are added and immobilized by the biotin. The presence of antibody in the test serum results in a yellow color when the substrate is added.

Ionescu-Matia *et al.* (1983) developed two micro solid-phase enzyme immunoassays for the detection of anti-HBs antibodies to improve detectability and to reduce the high cost of the commercial kit and the large volume of material required for the latter. In their test, HBsAg adw was diluted in 0.3 *M* carbonate buffer, pH 9.5, to a protein concentration of 4 μg/ml. Of this solution 50 μl was brought into the wells of rigid polystyrene plates and left for 22 hr at 4°C. The wells were then postcoated with 0.5% gelatin in 0.05 *M* PHS, pH 7.2.

Conjugates prepared with SPDP, which we have found to have questionable stability, were, nevertheless, able to detect anti-HBs antibodies at titers 625 times higher (geometric mean) than those of the commercial kit in the modified design.

Caution is warranted since various components of normal human serum may be associated with the 22-nm HBsAg particles (Schuurs and Wolters, 1975). Many experimental anti-HBs antisera may thus cross-react with normal human serum components. This could lead to a nonspecific rise in the titer. This nonspecificity can be tested or reduced by the inclusion of 10% normal human serum in the antiserum (the normal components would then neutralize the nonspecific antibodies).

The HBsAg can be detected by the method described by Wei *et al.* (1977) and Wolters *et al.* (1976). Anti-HBs antibodies from guinea pigs are immobilized on a solid phase. HBsAg present in the serum will be bound by these antibodies. Peroxidase-labeled anti-HBsAg antibodies may then detect the immobilized antigen. An improved sandwich enzyme-linked immunosorbent assay (ELISA) for hepatitis B virus surface antigen was developed recently using monoclonal anti-HBs for the solid phase and horseradish peroxidase-labeled sheep anti-HBs (Wolters *et al.*, 1985). The sensitivity of this method is 0.3 U/ml HBsAg, in the standard test procedure with two incubation steps of 60 min at 37°C. A slightly reduced sensitivity (0.5 U/ml) was noted when the two incubations were combined in a one-step incubation of 60 min at 37°C. The test for HBe is essentially the same (antibody–antigen–antibody sandwich).

The core antigen (HBcAg) appears in serum somewhat later than HBsAg but persists after the latter has disappeared and can, therefore, be the only marker of HBV infection until the subsequent appearance of anti-HBs antibodies.

ELISA procedures for HBV and its antibodies licensed in North America are Cordia H (Cordis Corp) and Auszyme (Abbott Laboratories) whereas in Europe Hepanostika (Organon, Oss) and Enzygnost Anti-HBs EIA (Hoechst-Behring) are frequently used. Cordis uses disks, whereas Abbott uses beads as the solid phase and Organon and Hoechst-Behring use microtiter plates. Cordis and Behring use alkaline phosphatase as the label, whereas Organon and Abbott use peroxidase. Alkaline phosphatase has superior stability and is less sensitive to water contaminants, but is less sensitive than peroxidase.

The Abbott Laboratories enzyme immunoassay test Auszyme II for the detection of hepatitis B surface antigen is described in Table 10.5.

c. Delta Antigen (Hepatitis D Virus)

The delta antigen (Rizzetto *et al.*, 1977; Rizzetto, 1983) seems to have a higher incidence in patients with fulminant hepatitis (Raimondo *et al.*, 1983) although an acute, simultaneous delta/hepatitis infection does not seem to increase the severity (Shattock and Fielding, 1983). The delta antigen or hepatitis D virus has the appearance of a large HBsAg particle (35 nm), and has HBsAg coat which surrounds an internal component consisting of a small RNA molecule (500 kDa) and "delta antigen." It can be detected in the nucleus of hepatocytes of patients late in the incubation period and in the beginning of the acute phase (Mushahwar *et al.*, 1984).

Shattock and Morgan (1984) developed an enzyme immunoassay to detect delta antigen and anti-delta antibodies. They designed a method with superior sensitivity. Instead of the usual antigen preparation from sonicated liver nuclei preparations, they derived the antigen from detergent-treated serum. Delta antigen is usually transient in serum (Smedile *et al.*, 1983), but delta antigenemia may be more frequent than previously suspected in drug abusers.

d. Non-A, Non-B Hepatitis Virus (HNANB)

HNANB has been reviewed extensively by Gerety (1981), and seems to consist of two distinct agents. Though non-A, non-B hepatitis is generally rather mild, a high rate of patients develop chronic liver damage (Berman *et al.*, 1979) and in some cases to a fulminant course leading to death in hepatic coma shortly after the onset of the disease.

Duermeyer *et al.* (1983) developed an ELISA to detect an antigen or

TABLE 10.5

Enzyme Immunoassay Test for Detection of Hepatitis B Surface Antigen (HBsAg) in Serum or Plasma (Auszyme II Procedure[a])

Reagents

1. Antibody to hepatitis B surface antigen (guinea pig)-coated beads
2. Antibody to hepatitis B surface antigen (goat): peroxidase (horseradish) conjugate concentration not less than 0.2 μg/ml in Tris buffer with protein stabilizers and preservative (Thimerosal)
3. Positive control (human HBsAg 6 ± 2 ng/ml in Tris buffer with protein stabilizers and 0.1% of sodium azide as preservative)
4. Negative control (recalcified human plasma, nonreactive for HBsAg and anti-HBs and preservative—0.1% sodium azide)
5. OPD (*o*-phenylenediamine–2 HCl) tablets
6. Diluent for OPD: citrate-phosphate buffer containing hydrogen peroxide

Procedure for Auszyme II test

Abbott Laboratories suggest three procedures for the detection of HBsAg in serum or plasma of patients:

	First incubation		Second incubation		Third incubation	
Procedure	Temp. (°C)	Time	Temp. (°C)	Time	Temp. (°C)	Time
A	38–41	120 min	38–41	60 min	15–30	15 min
B	15–30	16 hr	38–41	60 min	15–30	30 min
C	38–41	30 min	38–41	30 min	15–30	30 min

Three negative and two positive controls must be tested with each run of unknowns. All reaction trays containing controls and/or unknowns are subjected to the same process and incubation times. To avoid cross-contamination a separate disposable tip has to be used for each transfer

1. Bring all reagents to 15–30°C before beginning the assay procedure and swirl gently before using
2. Identify the reaction tray wells for each specimen of control date sheet
3. Dispense 200 μl of the negative control and positive control and 200 μl of the specimens into assigned wells
4. Dispense one coated bead into each well
5. Apply cover sealer to each tray. Gently tap trays to ensure that each bead is covered with the sample and that any air bubbles are released
6. For procedure:
   A: incubate the trays in the 38–41°C water bath for 120 min
   B: incubate the trays at 15–30°C for 16 hr on a level surface
   C: incubate the trays at 38–41°C water bath for 30 min

(*continued*)

TABLE 10.5 (*Continued*)

7. At the end of the incubation period, remove and discard the cover sealers. Aspirate the contents of the wells and wash each bead
8. Pipet 200 μl antibody to hepatitis B surface antigen (goat): peroxidase conjugate into each well
9. Apply cover sealer to each tray. Gently tap trays to ensure that each bead is covered with the conjugate and that any air bubbles are released
10. Procedures A and B: incubate the trays in the 38–41°C water bath for 60 min. Procedure C: incubate the trays in the 38–41°C water bath for 30 min
11. During the last 5 to 10 min of the incubation or after the final wash (step 12), prepare OPD substrate solution
12. At the end of the 60-min incubation period, remove the trays from the water bath. Remove and discard the cover sealers. Aspirate the contents of the wells and wash each bead. Remove all excess water from the top of tray by aspiration or blotting
13. Immediately transfer beads from wells to properly identified assay tubes. Align inverted rack of oriented tubes over the reaction tray, press tubes tightly over wells, then invert tray and tubes together so that beads fall into corresponding tubes. Blot excess water from the top of the tube rack
14. Pipet 300 μl of the freshly prepared OPD substrate solution into each tube containing a bead and into two empty tubes (substrate blanks)
15. Incubate tubes at 15–30°C for 30 min. To prevent accidental liquid spills into the tubes, cover box until incubation is complete
16. After the 30-min incubation, stop the enzyme reaction by adding 1 ml of acid to each tube containing a bead and to each of the substrate blanks. Do not allow acid solution to come into contact with metal. Air bubbles should be removed prior to reading absorbance
17. Select a wavelength of 492 nm on a spectrophotometer or use the Quantum analyzer. Blank the instrument by using one of the two substrate blank tubes. Read negative and positive controls, then process the unknowns. Visually inspect both blanks and discard those that are contaminated (indicated by yellow-orange color). If both blanks are contaminated, the entire run must be repeated

Note: following the reading of any specimen with absorbance greater than 0.5, cuvettes should be thoroughly rinsed with distilled or deionized water prior to reading the next tube. All absorbance values should be determined within 2 hr after addition of acid (1 *N* H_2SO_4)

18. Results: Cut-off value determination. The presence or absence of HBsAg is determined by relating the absorbance of the unknown sample to the cut-off value. The cut-off value is the absorbance of the negative control mean plus the factor 0.05 for procedures A or B or the factor 0.025 for procedure C, respectively. Unknown samples whose absorbance is greater than or equal to the cut-off value established with the negative control are to be considered reactive for HBsAg

Caution: Do not splash specimen or reagents outside of well or high up on well rim as it will not be removed in subsequent washings and may be transferred to the tubes causing test interference

[a] Abbott Laboratories.

antibodies to this agent. They found low frequencies in the low-risk group (normal blood donors), but higher among, e.g., patients with hemophilia and in prostitutes (no drug addicts). The latter finding may confirm the finding of Szumess *et al.* (1975) that non-A, non-B hepatitis virus is transmitted sexually.

10. AIDS-Related HTLV Virus (Retrovirus)

Litton Bionetics Laboratory has developed an ELISA test based on magnetic beads and intended for the detection of anti-HTLV antibodies.

a. Outline HTLV Procedure*

Principle of test: specific antigen from HTLV (i.e., p-24) is bound to the polycarbonate surface of beads of ferrous metal and supplied as such by Litton Bionetics Laboratory. These beads are transferred with a magnetic device from solutions containing serum, anti-IgG–enzyme conjugates, and substrate. Antibody present in the sample will be bound to the antigen on the bead and will in turn adsorb conjugate from the solution. The appearance of color once the bead is placed in the substrate solution indicates the presence of antibody. Materials required for this test are available from Litton Bionetics in a kit-form. Materials not supplied in these kits are the magnetic Transfer Device and Bead Dispensers (also available from Litton Bionetics).

b. Method

1. Sera to be tested and control sera (negative and positive) are diluted 50 times (5 μl/250 μl PBS–Tween-20) and dispensed (0.25 ml) into the designated well of a round-bottom 96-well plate.
2. In another plate, antigen beads are placed using a bead dispenser. The beads are then transferred simultaneously with the Magnetic Transfer Device into the wells with the sera.
3. The plate containing sera and beads is incubated for 45 min at 37°C with gentle rocking.
4. Transfer the beads to wash solution and raise and lower the magnet 12 times for efficient release of nonspecifically adsorbed material.
5. Transfer the beads to conjugate-containing (0.2 ml) wells and incubate for 45 min at 37°C.

*Procedure for the HTLV Bio-EnzaBead test kit provided by Litton Bionetics Laboratory Products Division.

6. Transfer the beads sequentially to two plates containing wash solutions (move the beads, respectively, two and four times up and down in the wash solutions).

7. Transfer the beads to a clean plate containing substrate (in the Bionetics HTLV Bio-EnzaBead Test kit, ABTS since peroxidase is used). Arrest the reaction after 10 min.

8. Remove the beads and read the results on a white background or with a test reading mirror (with a white paper on top of the plate).

c. *Results*

Negative control sera should have, at most, negligible coloration after the usual incubation period of 10 min. The positive control serum should yield a definite but light green color (1 to 2+). When the color in the well with the test sera is essentially equal to or darker than those with the positive control sera, they are considered positive.

The acquired immune deficiency syndrome (AIDS) may be spread perinatally, parentally, and sexually. The AIDS is transmitted frequently by blood transfusions and by blood products from donors who have been infected with the T lymphotropic retrovirus type III (HTLV-III) designated also as the lymphadenopathy-associated virus (LAV). Recently Neurath *et al.* (1985) described an enzyme-linked immunoassay (ELISA)-inhibition tests to detect antibodies directed against LAV antigens (P24 and PL proteins). The application of their tests shows that among individuals belonging to groups at high risk of developing AIDS, 45.5% were positive for anti-PL and 42% for anti-P24, respectively. The authors suggested the development of a test measuring antibodies to all viral antigens as the most suitable solution for LAV/HTLV-III diagnosis.

11. Enzyme Immunoassays of Arboviruses

The applicability of ELISA and immunoperoxidase techniques to the study of arboviruses has been demonstrated by this laboratory (Kurstak *et al.*, 1980; Charpentier *et al.*, 1982). McLean and his colleagues (1978, 1979) could demonstrate the intracytoplasmic replication of California encephalitis (CE) virus in both domestic and wild mosquitoes from arctic regions with the immunoperoxidase technique. Enzyme immunoassays have been applied for flaviviruses (Catanzaro *et al.*, 1974), bunyaviruses (McLean *et al.*, 1978), and alphaviruses (Stollar *et al.*, 1979).

Immunoperoxidase procedures also offer the possibility of comparing at

Fig. 10.9. Detection of antigens of the viral envelope of Chikungunya virus in infected BHK-21 cells. Electron microscopy indirect immunoperoxidase staining.

light and electron microscope levels the processing and localization of the viral antigens. We have followed the infection sequences of different alphaviruses (Sindbis, Chikungunya) by immunoperoxidase. For example, BHK-21 cells infected with Chikungunya virus contained, as revealed by the immunoperoxidase technique, antigens of the viral envelope in the cytoplasmic membrane (Fig. 10.9), and on some intracytoplasmic vacuoles. Kurstak *et al.*, (1980) and Charpentier *et al.* (1982) postulated, therefore, a process of reverse pinocytosis or exocytosis since specifically stained vacuolar membranes were incorporated in the cytoplasmic membrane. The envelope antigens are incorporated by budding at areas of the cytoplasmic membrane which were positively stained. The usefulness of the immunoperoxidase technique has also been shown for the study of the synthesis of antigens of Dengue-2 and Sindbis viruses. With this method it was possible to demonstrate that Dengue-2 virus antigen is inserted into the plasma membranes (Catanzaro *et al.*, 1974) and that Sindbis virus core antigen is easily detected in the cytoplasm of CER cells (Kurstak *et al.*, 1980).

In the study with peroxidase-conjugated *Staphylococcus aureus* protein A used for indirect staining of arbovirus antigens in mosquito cell lines, Zhang Yong-He *et al.* (1984) demonstrated that their method detects infection 2 days before the appearance of the cytopathic effect. Konishi and Yamaoka (1983) proposed a rapid enzyme-linked immunosorbent assay of whole blood for detection of antibodies to Japanese encephalitis virus. The magnetic processing ELISA system for large-scale epidemiological studies permitted them to save a total of 1 hr in comparison with the ordinary ELISA procedure. This method seems to be useful for the epidemiological studies of arbovirus infections in the field.

The detection of Japanese encephalitis virus antibody in human and a variety of animal sera using biotin-labeled sandwich ELISA was proposed also by Chang *et al.* (1984). Kuno *et al.* (1985) developed a simple antigen capture enzyme immunosorbent assay (AgC-ELISA) for identifying Dengue virus. The method used serotype-specific monoclonal antibodies as capture antibodies and an enzyme conjugate of a flavivirus-reactive monoclonal antibody as a detecting antibody. The Dengue virus strains, representing all 4 serotypes isolated from various parts of the tropics, were identified by this AgC-ELISA method. In a recent study Bundo and Igarashi (1985) developed an antibody-capture ELISA (AbC-ELISA) method for detection of IgM antibodies in sera from Japanese encephalitis and Dengue hemorrhagic fever patients.

11

Automation in Enzyme Immunoassays and Commercial Kits

Solid-phase enzyme immunoassays are often performed manually. For their performance on a large scale, mechanization and automation are necessary.

Ruitenberg *et al.* (1976) and Ruitenberg and Brosi (1978) described an on-line system for macroELISA (in polystyrene tubes). Microplate enzyme immunoassays have also been automated. Homogeneous enzyme immunoassays lend themselves to automation and are performed in kinetic analyzers or similar equipment available in clinical chemistry laboratories.

A. TOTAL AUTOMATION

Total automation for the determination of antibody levels is possible with the Auto Analyzer II (produced by Technicon Instruments Corporation, Tarrytown, New York). This instrument enables the processing of 30–300 samples per hour. It consists of a rotary carousel sampler tray with aspirator for the introduction of samples, a segmented proportioning air pump to carry samples and add reagents, timed incubation coils and incubation samplers, colorimetric readers, and a recorder with computerized digital output.

B. SEMIAUTOMATION

For most purposes, semiautomated systems are the best compromise between cost and efficiency (Sever, 1983).

1. Samplers and Dispensers

Placement of undiluted specimens in cuvettes in blocks or carousel trays can be achieved with samplers offered by Technicon Instruments Corporation. Dispensing constant volumes of reagents can be rapidly performed with semiautomated or fully automated instruments (e.g., Dynatrop SR from Dynatech Laboratories, Alexandria, Virginia, and the Wash Dispenser from M.A. Bioproducts, Walkersville, Maryland). The Dynadrop SR dispenser (Fig. 11.1) delivers volumes of 25 or 50 μl, whereas the Wash dispenser has larger capabilities. Both instruments will deliver fluid for 1 column (i.e., 8 tips) whereas the 96-channel Autopipette of Dynatech is capable of dispensing simultaneously 96 volumes of 25 or 50 μl into a plate. The Dynadrop MR (Fig. 11.2) is completely automated (12 tips). The Titertek Autodrop (Fig. 11.3) is also completely automated but volumes can be selected in the range of 20–300 μl.

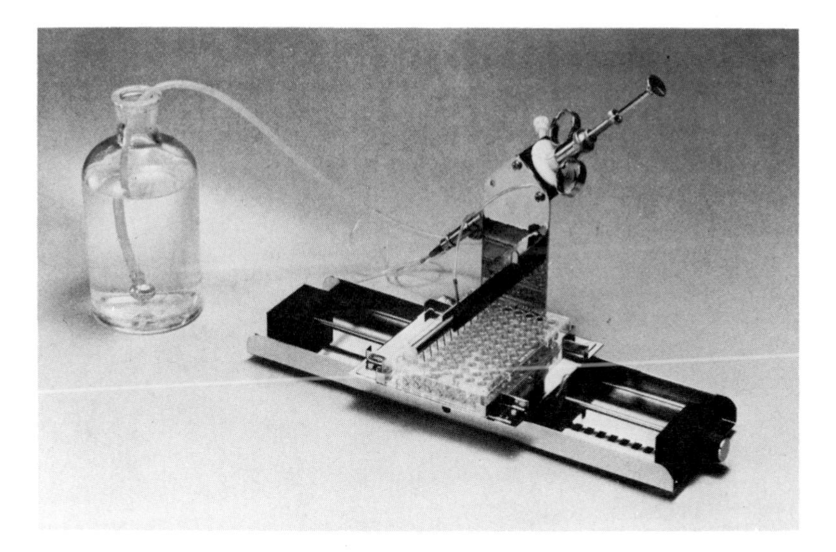

Fig. 11.1. The Dynadrop SR dispenser (Dynatech Laboratories, Alexandria, Virginia).

Fig. 11.2. The Dynadrop MR (from Dynatech Laboratories). A fully automated pumping dispenser with 12 tips.

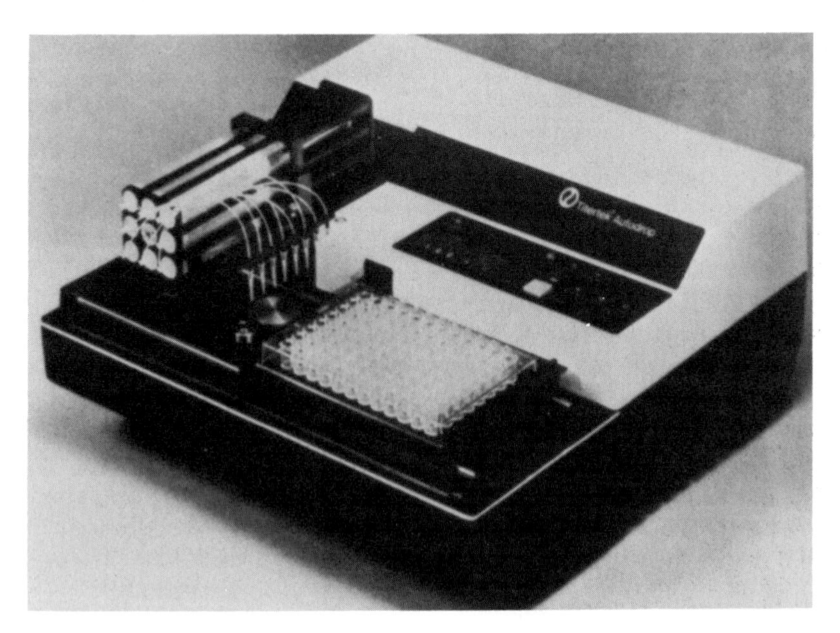

Fig. 11.3. Automated microprocessor-controlled dispenser—Titertek Autodrop (EF-LAB OY, Helsinki, Finland, and Flow Laboratories, McLean, Virginia).

2. Diluters

Titertek (Fig. 11.4) offers 4-, 8-, and 12-channel configurations of hand-held diluters in the range of either 5–50 or 50–200 μl. An instrument based on the "tulip"-loop configuration is offered by Dynatech (for 25 or 50 μl).

3. Washers

A variety of semiautomated and automated washers for microplates are available. The Miniwash of Dynatech Laboratories is a useful hand-held

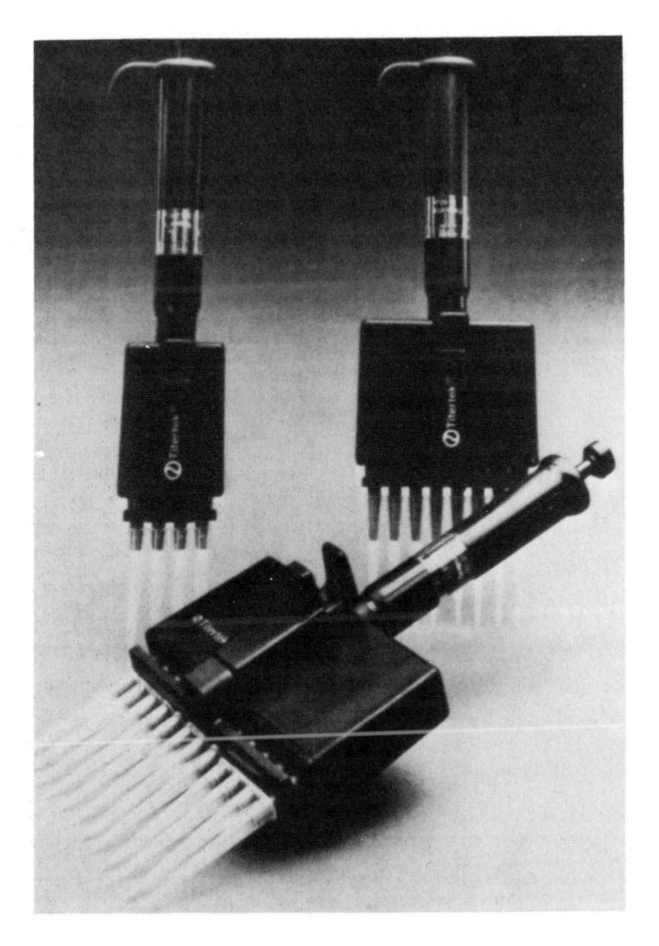

Fig. 11.4. The Titertek (EFLAB OY, Helsinki) diluters.

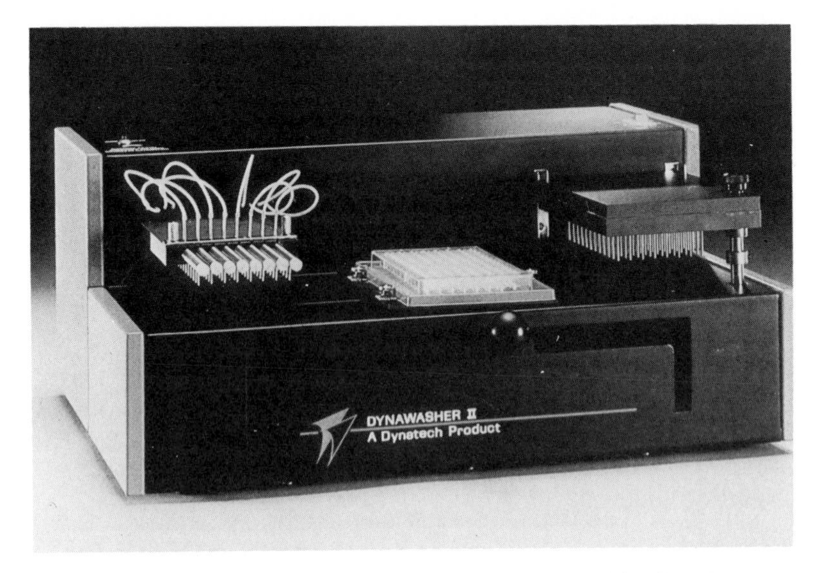

Fig. 11.5. The Dynawasher II (Dynatech Laboratories), a 96-well washer.

washer-aspirator. The Dynawasher II (Fig. 11.5) is semiautomated, dispenses wash fluid, and aspirates by movement of the plate. The Multiwash from Titertek (or Flow Laboratories) is fully automated.

4. Readers

Most ELISA readers can be used with microplates. The wavelength used is obtained with the appropriate filters (though they may not be optimal in certain situations).

Ackerman and Kelly (1983) constructed a photometer equipped with a fiber optic probe (Brinkman Microtitre Plate Reader). The light sent through the well by a probe tip using fiber optics is reflected back to a photocell. Phase-shifted modulating light is generated by an alternating current and returned (through the liquid) to another light guide in the same probe. An electronic chopper synchronized with the light source negates the effects of extraneous or ambient light (Fig. 11.6).

The Beckman DU-8 UV/VIS Computing Spectrophotometer can also be equipped with a Microplate Analyzer Accessory. This instrument can use dual wavelengths (Fig. 11.7).

The Dynatech Minireader II (Fig. 11.8) can be used with a wide range of

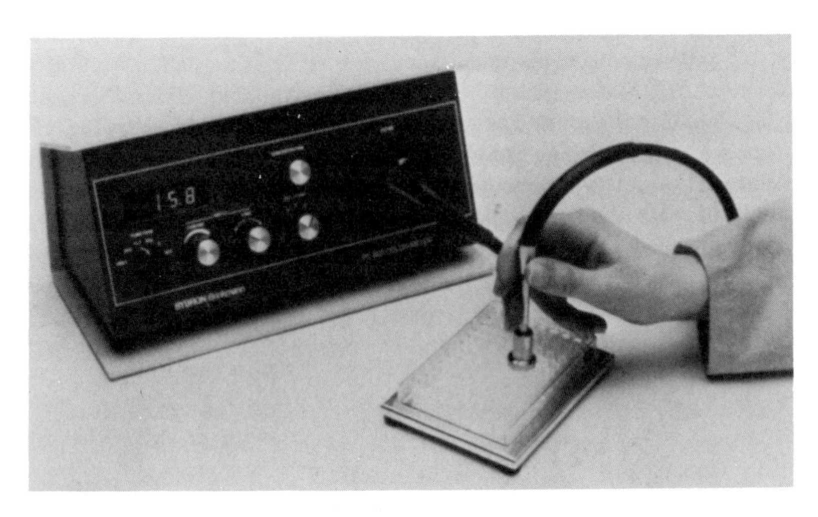

Fig. 11.6. Microtiter Plate Reader for ELISA from Brinkman Instruments (Westbury, New York).

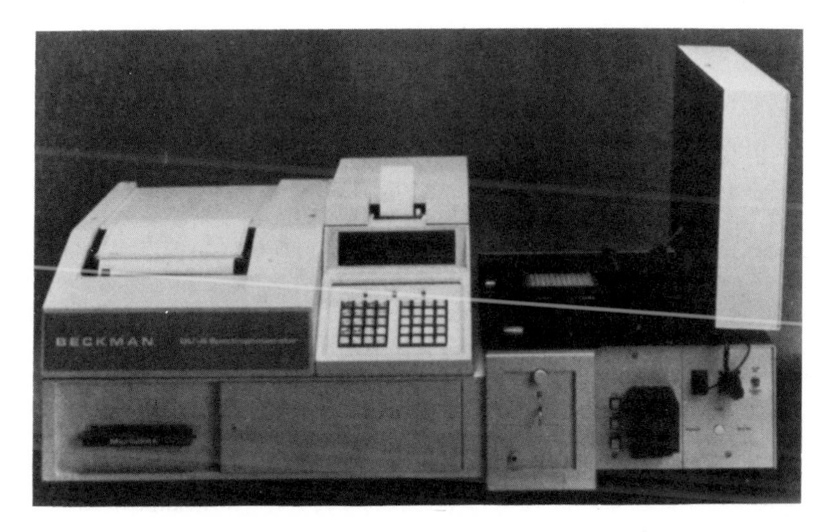

Fig. 11.7. The DU-8 Microplate Analyzer—attachment to the DU-8 UV/VIS Computing Spectrophotometer for ELISA from Beckman Instruments Inc. (Irvine, California).

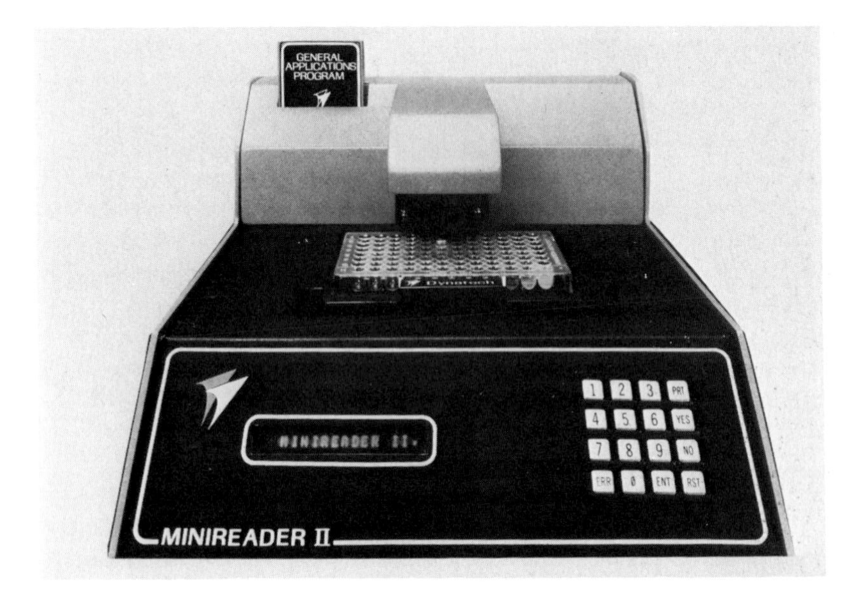

Fig. 11.8. The Dynatech Minireader II for a 96-well plate (Alexandria, Virginia).

microelisa plates. Relatively inexpensive is the MicroELISA Minireader (MR 590) from Dynatech (Fig. 11.9). The plate is moved manually. The Microplate Reader (MR 600) from Dynatech has dual wavelength, can be interfaced with a computer (Apple), and has automatic plate movement (Fig. 11.10).

The Bionetics Autoreader (LB1-321 is specifically designed for the Bio-Enza Bead system (Litton Bionetics, Kensington, Maryland). An interfaced computer can analyze and print the results.

The Titertek Multiskan and the Bio-Rad ELISA reader measure 8 wells simultaneously and print the results; interfacing with a computer is possible.

The Quantum II from Abbott Laboratories is a dual beam reader for single-tube ELISA [i.e., for tests using beads (Rotazyme, etc.)]. This model is a preprogrammed reader which plots value and gives percentage activity (Fig. 11.11).

The Virion reader (Institut Virion AG, Switzerland) simultaneously displays all 96-well results on a 12-in. screen (Fig. 11.12) before it makes hard copy of the data. Other readers include the Behring (West Germany) Elisa Processor M (Fig. 11.13), the Organon/Teknika (Belgium) Micro-

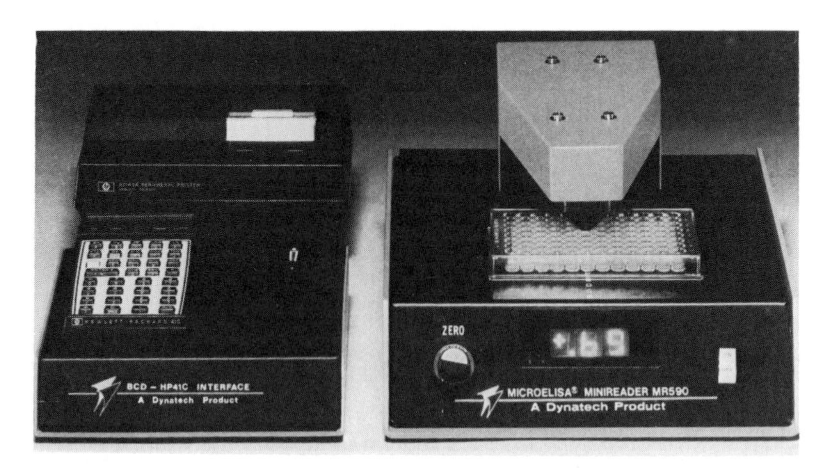

Fig. 11.9. MicroELISA Mini Reader MR590 from Dynatech Laboratories (Alexandria, Virginia).

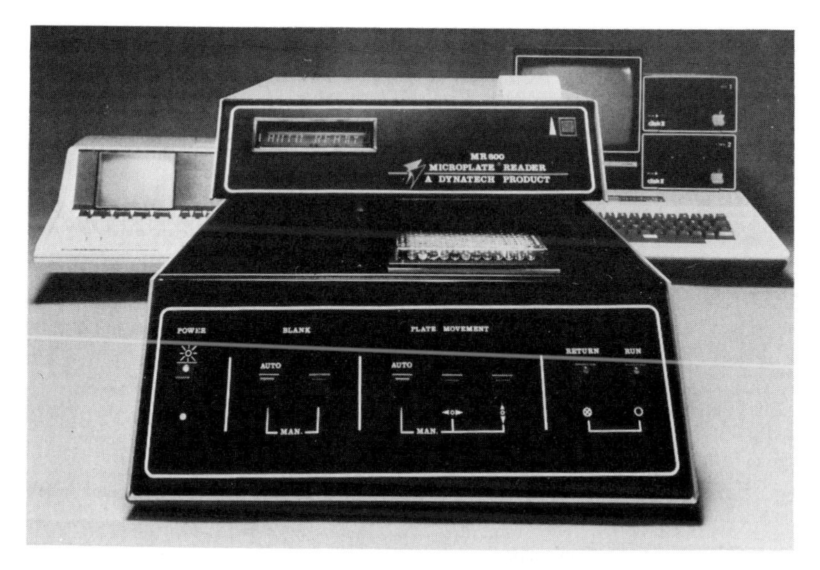

Fig. 11.10. The Dynatech Microplate Reader MR 600—a dual wavelength reader and printer.

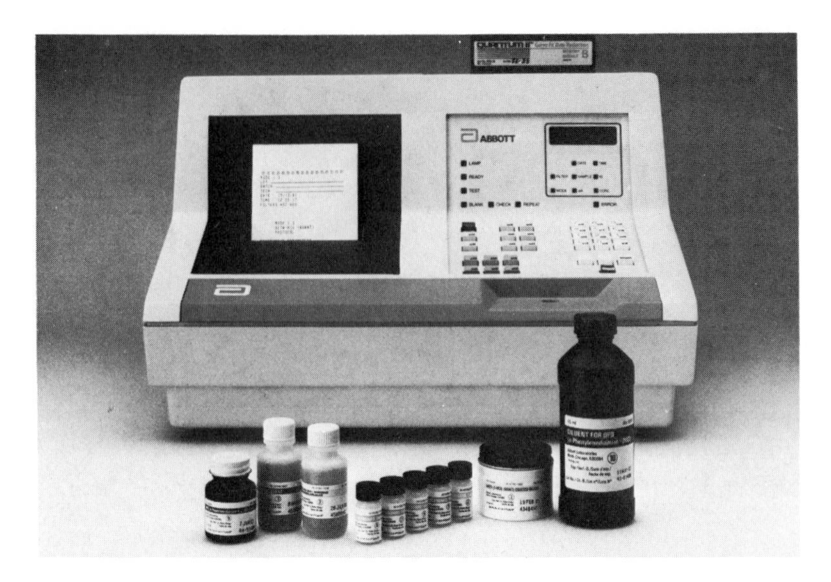

Fig. 11.11. Quantum II—programmed reader for single-tube ELISA (Abbott Laboratories, North Chicago, Illinois).

Fig. 11.12. Virion reader from Institut Virion AG (Zürich, Switzerland)—customizable high-speed fully integrated data processing capabilities for photometric enzyme immunoassay measurements.

Fig. 11.13. Behring ELISA-Processor M (Behringwerke AG, Marburg, West Germany).

Elisa system (Fig. 11.14), the Boehringer (West Germany) Elisa automated test "Enzymum-test system ES22" in coated tubes (Fig. 11.15), the Artek Model 210 (Artek Systems Corporation, Farmingdale, New York), the Gilford Automated EIA system (For cuvette-packs; Gilford Company, Oberlin, Ohio), and the Bio-Rad System (for microplates; Bio-Rad, Richmond, California).

5. Blotting

Bio-Rad offers a Bio-Dot apparatus for immunodotting applications. A sheet of nitrocellulose is inserted between a 96-well sample template and a support plate of the vacuum reservoir. Washing fluid is drawn through the membrane to the vacuum. Problems may arise with dense fluids (blocking of the membrane).

C. COMMERCIAL KITS FOR ENZYME IMMUNOASSAYS

A most useful source for finding commercial suppliers of monoclonals, antisera or serum fractions, enzymes, and other useful reagents for enzyme

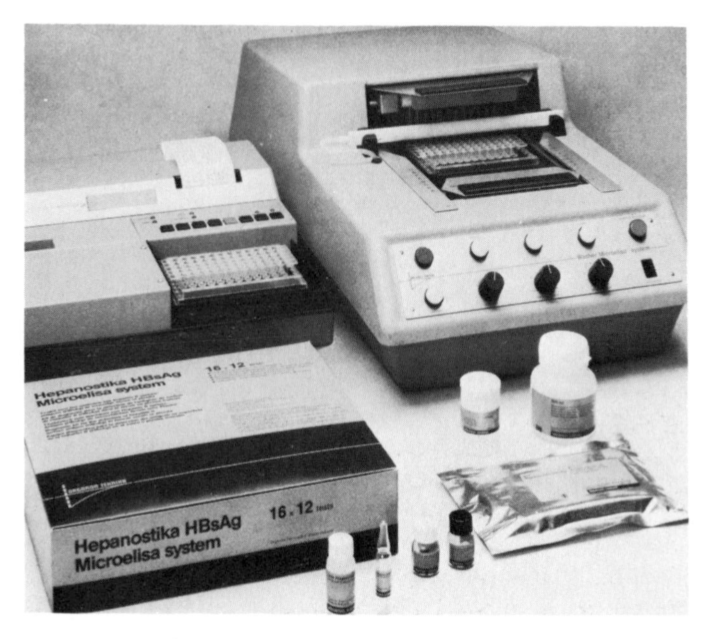

Fig. 11.14. Organon/Teknika MicroElisa system (Turnhout, Belgium).

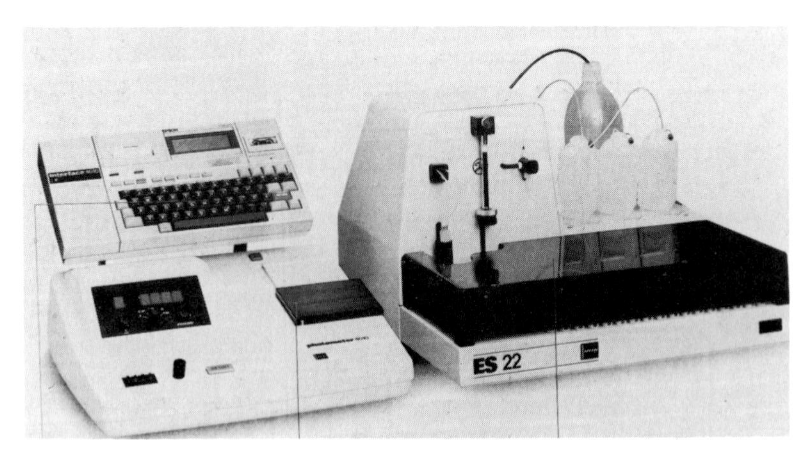

Fig. 11.15. Automated Enzymum-test system ES 22 from Boehringer (Mannheim, West Germany).

immunoassays has been prepared by Linscott (Linscott's Directory of Immunological and Biological Reagents, 40 Glen Drive, Mill Valley, California 94941).

1. Immunohistochemical Staining

A staining service for human antigens is offered by Immuno-Mycologics Inc. (POB 1151, Norma, Oklahoma).

A large number of immunoperoxidase kits are available, particularly for human antigens. Most are based on the PAP system though some are based on avidin–biotin. An increasing number of these kits are based on monoclonal antibodies.

Companies offering immunohistochemical staining kits are Milab (POB 20047, Malmö, Sweden) which offers PAP kits for the detection of a wide range of human proteins; Becton and Dickenson (Monoclonal Antibody Center, 2375 Garcia Ave., Mountain View, California) offering monoclonal antibody kits, particularly for T cell surface antigens; Enzo (Biochem Inc., 325 Hudson Street, New York, New York) which supplies PAP kits for species of origin tissue identification and reverse transcriptase (AMV); Vector Laboratories Inc. (1429 Rollins Rd, Burlingame, California) offering avidin and biotinylated immunoglobulins as well as custom biotin labeling of proteins; Biogenex Laboratories (6529 Sierra Lane, Dublin, California) offers monoclonal antibody kits to detect tumor antigens; Dako Corp (22 North Molpas Street, Santa Barbara, California) and Accurate Chemical Scientific Corp (300 Shames Dr., Westbury, New York) offer PAP kits for the detection of hormones, tumor antigens, immunoglobulins, and some viruses (hepatitis B, herpes, papilloma); Ortho Diagnostic Systems Inc. (Raritan, New Jersey) offers a wide range of "Histoset" immunoperoxidase staining kits (on the basis of the PAP technique) for immunoglobulins, enzymes, oncofetal antigens, tissue specific markers, and other cellular antigens, some viruses (herpes, hepatitis B), and hormones.

2. Enzyme Immunoassay Kits

The number of commercial kits is rapidly expanding and any review of this area is outdated almost instantaneously. Therefore, emphasis is given on the orientation of the different companies and not on particular products. For the same reason, whether the kits are intended to be used for the detection of antigens or antibodies has not been specified.

The following companies were at the beginning of 1984 actively offering enzyme immunoassay kits: Accurate Chemical and Scientific Corp. (300 Shames Dr., Westbury, New York) supplies kits for oncofetal antigens, immunoglobulins, and viruses; Abbott Diagnostics (North Chicago, Illinois) is primarily involved with viruses (particularly hepatitis and rotaviruses), oncofetal antigens, hormones, and bacteria (*Neisseria gonorrhoeae*); their kits are based on the bead system; Biomedical Technologies (22 Thorndike Street, Cambridge, Massachusetts) is involved in custom-assay development and in human protein kits; BioMérieux (Marcy-l'Etoile, Charbonnières-les-Bains, France) is among others supplying kits for viruses and immunoglobulins and viruses; BRL-Life Technologies (POB 6009, Gaithersburg, Maryland) offers a kit for the detection of terminal deonynucleotidyltransferase (see Chapter 9); Cordis Laboratories (POB 523580, Miami, Florida) markets a large array of kits for the detection of (or of antibodies to) viruses, bacteria, parasites, DNA, heterophile antibody, immune complexes, rheumatoid factor; Calbiochem-Behring (POB 12087, San Diego, California) is particularly involved with kits pertaining to viruses and immunoglobulins; Dako Corp (22 North Milpas Street, Santa Barbara, California) is offering kits for oncofetal antigens and viruses (see also immunohistochemical kits); Labsystems Inc. (POB 48723, Chicago, Illinois) offers a kit to detect tetanus toxoid IgG antibody; M. A. Bioproducts (Bldg 100, 100 Biggs Ford Rd., Walkersville, Maryland) in addition to kits for viruses and hormones, also offers an anti-chlamydia kit; Litton Bionetics Inc. (2020 Bridge View Dr., Charleston, South Carolina) has developed their own distinctive kits for viruses (recently also for AIDS-associated virus), bacteria, and parasites. A very active supplier of homogeneous immunoassays for drugs or haptens is Syva Company (POB 10058, Palo Alto, California). Other companies offering some kits are Monoclonal Antibodies Inc. (2319 Charleston Rd., Mountain View, California), BBL Microbiology Systems (POB 243, Cockeysville, Maryland), Diagnostic Products Corporation (5700 West 96th Street, Los Angeles, California), Difco Laboratories (POB 1058, Detroit, Michigan), ICL Scientific (18249 Euclid Street, Fountain Valley, California), New England Immunology Associates (53 Smith Place, Cambridge, Massachusetts), and Pharmacia Diagnostics (800 Centennial Ave., Piscataway, New Jersey).

12

Screening of Recombinant DNA Expression Libraries

It is necessary to distinguish recombinant DNA molecules from non-recombinant or aberrant recombinant molecules after ligation of the DNA fragments and transformation of the host. If strategies are chosen that require the recombination to take place for successful transformation (e.g., alkaline phosphatase treatment and A/T or G/C tailing) they need not be correct. Different methods are available for this purpose. The identification of altered phenotypes due to the recombinant plasmid is an indirect method which can be used with certain recombinants. Direct identification methods such as DNA sequencing, or using DNA to select a specific mRNA for *in vitro* translation, or a combination of these methods are time consuming and not convenient for the screening of large libraries. The *in situ* hybridization of plasmid DNA to a complementary radiolabeled probe (Tijssen and Kurstak, 1977), or immunological screening, may become very important for large cDNA expression and replace most current methods since it is not necessary to enrich for the clone of interest (physical fractionation such as gel electrophoresis, or immunoprecipitation of polysomes). However, these immunoassays depend on the availability of an antibody directed against the protein of interest and on the faithfulness with which the cloned DNA fragment is transcribed and translated *in vivo*. This method is most suited for cDNA which, when cloned into plasmids that promote expression of the cDNA in the host, yields appropriate translation products.

A. CONSTRUCTION OF RECOMBINANT TRANSFORMANTS

The construction of a cDNA expression library is fairly standard (Goodman, 1979, Kurtz and Nicodemus, 1981; Rodriguez and Tait, 1983; Binns *et al.*, 1985). The cDNA should be cloned in the correct orientation with respect to the promoter and translational start site in the vector. Following the construction of the recombinant plasmids, the host (*E. coli*) is transformed with these plasmids, e.g., with the procedure developed by Hanahan (1983). The transformed *E. coli* are then plated on 82-mm nitrocellulose filters (Millipore Triton-free, HATF) overlaid on appropriate (e.g., ampicillin) plates to give 1000–2000 colonies per filter. The fewer the colonies per filter, the greater the number of filters needed for plating, but the greater the probability of isolating a single colony from a primary screen and thus avoid work necessary to obtain a pure clone. If, however, little antibody is available, or the antibody is precious, high-density plating is desired. Nevertheless, under many selective conditions, cells in the neighborhood of the transformant will grow once a colony of the transformant is established. This phenomenon, known as cross-feeding, may thus result in colonies consisting of a mixture of transformed and nontransformed cells. Single-colony isolates should, therefore, be prepared. Replica plating is carried out as described by Hanahan and Meselson (1980).

B. IMMUNOLOGICAL SCREENING OF LARGE cDNA EXPRESSION LIBRARIES

1. Chromogenic Detection of Antigen Producers in Bacteriophage Plaques in Agar Plates

Kaplan *et al.* (1981) have developed an enzyme immunoassay to screen antigen-positive phages among a large excess of negatives. Peroxidase-conjugated antibody, incorporated into a soft agar layer of a plaque assay system, is precipitated locally by the antigen produced during plaque formation. For this purpose, indicator cells (50 μl), phage (50 μl), and enzyme–antibody conjugate (20 μl) were mixed with 0.75 ml of 0.6% soft agar and poured in 5-cm plates containing 2.5 ml of 1.5% agar. After an incubation of 18 hr at 35°C, the plates were fixed with double-sided adhesive tape to the bottom of a plastic dishpan (8–10 dishes/pan), followed by a soaking (dialysis) in 2 liters PBS (twice for 10–24 hr). The enzyme was then detected by adding 4 ml of a solution containing DAB (0.5

mg/ml) and hydrogen peroxide (0.015%) for about 10 min. The sensitivity of this technique decreases considerably at high densities. However, this method eliminates intermediate steps such as autoradiography and alignment on an autoradiographic film with the original plate required in a widely used plastic overlay method (Broome and Gilbert, 1978). The peroxidase method described by Kaplan *et al.* (1981) allows the detection of antigen in a large number of plaques. High-density problems can be decreased for screening by using elevated concentrations of indicator cells. This also limits plaque size, and immunoprecipitate is confined to the same area, resulting in an intense staining. It also reduces the time required for plaque formation.

2. Solid-Phase Immunological Screening

Sandwich Procedures

The ability to detect translation products of cloned eukaryotic DNA fragments in bacteria is an important advance. Even fragments of a protein sequence, containing an epitope, can be detected by immunoassays. Broome and Gilbert (1978) devised a solid-phase sandwich assay for this purpose. They coated a plastic dish with the IgG fraction of an antiserum and pressed this dish onto the agar (without air bubbles), the colonies on which had been lysed by heat (2 hr at 42°C) and incubated for about 3 hr at 4°C. The dish was then removed, washed, and the antigen recovered by the solid-phase immobilized antibody detected with labeled antibodies (Fig. 12.1).

3. Direct Screening on Nitrocellulose

Hanahan and Meselson (1980) developed a method for plasmid screening at high colony density involving a high-density enrichment step and a low-density isolation step. In the enrichment step, up to 10^5 colony-forming bacteria are spread on a nitrocellulose filter laid on an agar plate and after small colonies have been established replicated to other nitrocellulose filters.

For 100-mm-diameter plates, 82-mm filters are used. The filters are floated on water, submerged when wet, sandwiched between dry Whatmen 3 MM filters, wrapped in aluminum foil, and autoclaved in liquid cycle. The pack of filters is kept in a plastic bag in order to maintain humidity.

For the immunological screening, bacteria (in colonies of about 2 mm)

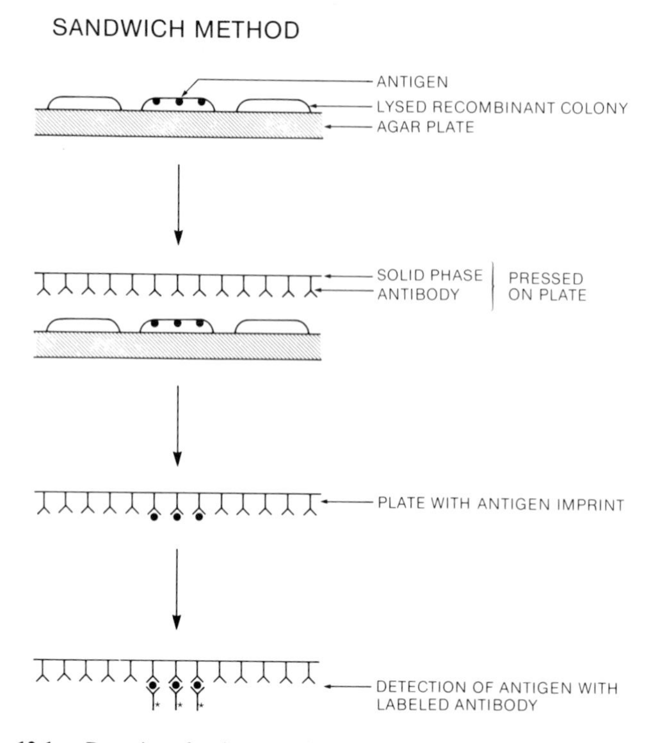

Fig. 12.1. Detection of antigens produced by recombinant DNA in colonies. The colonies are lysed to liberate the antigen. Subsequently, a solid phase coated with antibody is pressed on the plate to obtain an antigen imprint on this sensitized solid phase. The retained antigen can in turn be detected by immunoassays employed in the sandwich method. Though ^{125}I-labeled antibodies are still used on a large scale for this purpose, the highly sensitive enzyme probes have important advantages over radiolabels, and it is expected that they will replace the radiomarkers.

need to be lysed. For this purpose (Fig. 12.2) the nitrocellulose filters are removed from the media plates and suspended in a chloroform vapor chamber for about 20 min. The filters are then soaked in 0.05 *M* Tris–HCl buffer, pH 7.5, containing 0.15 *M* NaCl, 5 m*M* $MgCl_2$, 3% BSA, 1 μg DNase per ml, and 40 μg of lysozyme per ml and agitated gently overnight at room temperature. The bacterial debris is rinsed with TBS (0.05 *M* Tris–HCl buffer, pH 7.5, containing 0.15 *M* NaCl). The subsequent detection of positive colonies is performed as described in Chapter 4.

Helfman *et al.* (1983) preadsorb the primary antibody with bacterial lysates prepared from the nontransformed strain. The bacteria are concen-

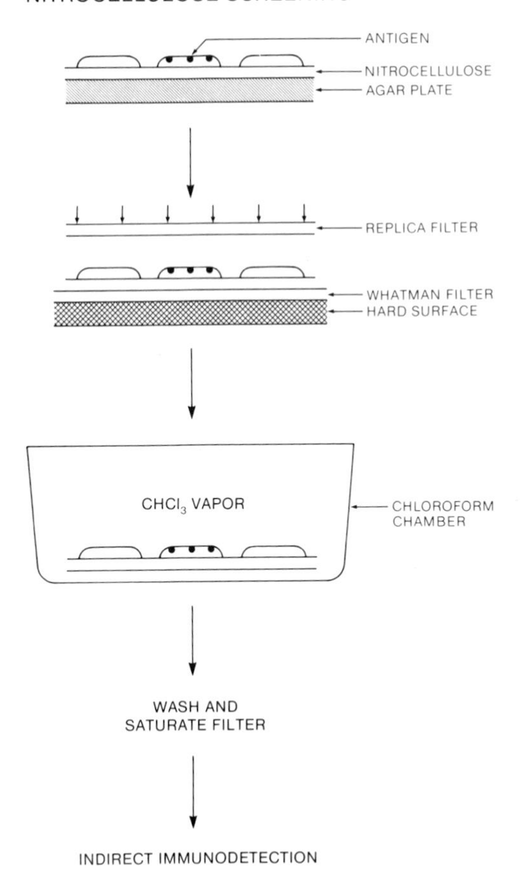

Fig. 12.2. Immunodetection for recombinant cDNA expression libraries. Bacteria are seeded on nitrocellulose filters on agar, and the colonies are replicated on several other filters on a hard surface. The filters can be punched with a needle for localization purposes. One of the filters, used for screenings, is incubated in a chloroform chamber to lyse the colonies. Bacterial debris is washed away, and the filter is saturated with an inert substance to prevent nonspecific adsorption of immunoreactants. The colonies expressing the antigen can then be identified by the indirect or other highly sensitive enzyme immunoassays.

trated 100-fold for this purpose by centrifugation and resuspension in deionized water, followed by boiling for 5–10 min. One milliliter of the lysate is added to 100 ml antiserum and incubated for 2 hr at 4°C. The antibody is then diluted in TBS so that a certain minimum amount (e.g., 1

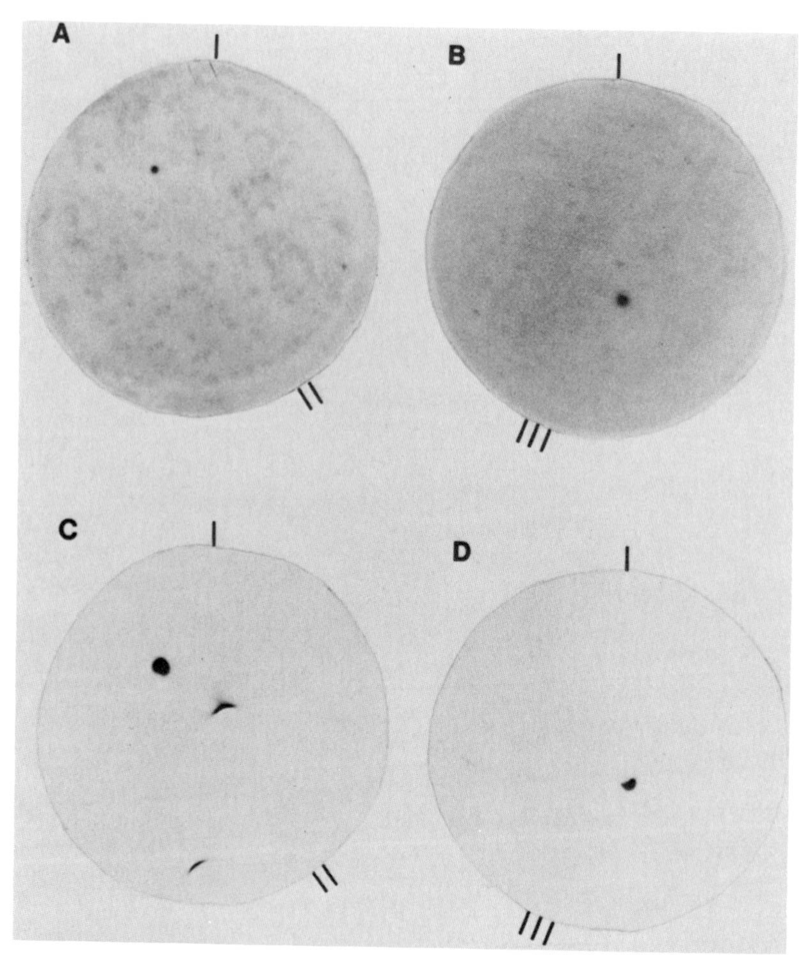

Fig. 12.3. Screening of recombinant DNA expression libraries. An incubation with antiserum, adsorbed with control bacterial lysates, and diluted in Tris-buffered saline, followed by an incubation with the conjugate allowed the detection of colonies with positive expression (A,B,C,D). Courtesy of Dr. D. M. Helfman with permission to reproduce.

ng) of purified antigen can be detected. Controls with purified antigen on nitrocellulose (Chapter 4) should be included. The optimum incubation periods and temperature depend on the nature of the antigen–antibody interaction (electrostatic, lower temperature; hydrophobic, higher temperature) and the avidity (low avidity, short incubation but at high concentration). Usually incubation periods are 1–2 hr (50–1000 times diluted antiserum), followed by five washes in TBS, an incubation with labeled antibody, washes, and enzyme detection (Chapter 4). Nonionic detergents at low concentration (0.05%) may decrease background staining.

A faint background can be beneficial since this will help locate the positive clone. Restreaking followed by dotting on a nitrocellulose grid can then help obtain a pure clone. Though for immunological screening ^{125}I-labeled antibody is generally used (Fig. 12.3), it can be expected that procedures based on enzyme markers will be increasingly important.

13

Vaccine Assessment and Standardization

Recent advances in immunology, biotechnology, and molecular biology have an accelerating effect on the development of vaccines. Significant progress has been made in three areas: (i) the development of vaccines against agents which hitherto could not be prevented, (ii) safer vaccines, and (iii) a much better cost/efficiency ratio.

Immunoassays are useful for different steps in the development of vaccines: (i) the antigenic characterization of the infectious agents (e.g., with monoclonal antibodies); and (ii) internal and external quality assessment schemes in research and clinical assays and in production.

A. SAFETY AND EFFICACY OF PROSPECTIVE VACCINES

All biological products to be used as vaccines or in vaccine preparations have to undergo rigorous safety tests and qualitative and quantitative standardization. Another particularly important aspect is that the cellular and humoral immune response should be measurable and quantified. Immunoassays can be used for this purpose and replace much more expensive *in vivo* procedures.

The standardization of monoclonal antibodies used for these purposes has to take into account the following factors (WHO, 1980):

1. The hybridoma clone used for the production of the monoclonal antibodies should be characterized. Its origin, its stability, the number of passages undergone, and the conditions of conservation should be specified.

2. The specificity of these antibodies should be well studied and described, preferably by using reference antigens.

3. The physicochemical characterization of the antibody should confirm its clonal origin. This characterization should be extended to the identification of the isotype, heavy and light chains, and the allotypes. Von Seefried *et al.* (1981) compared the radioimmunoassay and the enzyme immunoassay for *in vitro* potency testing of inactivated poliomyelitis vaccine. Peroxidase was used as the enzyme and *o*-phenylenediamine as the hydrogen donor. The results of both tests compared favorably in sensitivity and reproducibility with those from a complement fixation procedure. They were also able to compare vaccines from different manufacturers (results expressed in terms of D-antigen units). Antigen mass determination or potency testing *in vitro,* however, does not correlate well with immune responses *in vivo* (Salk *et al.*, 1978). Standardization of poliovirus antigens by animal (monkey) potency tests presents its own problems. Therefore, it is necessary to rely in large measure on a standardized manufacturing process to obtain consistently potent vaccines. The D-antigen induces neutralizing antibodies and is widely tested with the insensitive gel diffusion method. The enzyme immunoassays developed by van der Marel *et al.* (1981) are highly sensitive (approximately 1 DU/ml) and require very small amounts of material. They employed the indirect enzyme method successfully for D-antigen quantitation of $AlPO_4$ adsorbed DT- and DPT-polio vaccines. The limits for *in vivo* potency come very close to the limits for *in vitro* detection of antigen in enzyme immunoassays (Moynihan and Peterson, 1981). Blondel and Crainic (1981) adapted the immunoblotting technique (using peroxidase conjugates) for detection of the reaction between vaccine poliovirus structural polypeptides separated by electrophoresis and neutralizing anti-polio IgG.

The immune response to subviral fractions is also conveniently established by enzyme immunoassays. Mills *et al.* (1982) investigated the specificity of the rabbit antibody response to inoculation of subviral fractions of human cytomegalovirus using extracts of infected cells and then proceeded to examine the responses obtained after inoculation of envelope material. Similarly, Soulebot *et al.* (1982) used an enzyme immunoassay developed by Soula *et al.* (1981) for the detection of infectious bovine rhinotracheitis to establish the immune response after vaccination with viral subunits of this virus.

In some cases the evaluation of the immune status of vaccine recipients has been particularly difficult, e.g., for mumps for which a simple and sensitive method has been lacking (Kuno-Sakai *et al.*, 1984). The ODs of ELISA of postvaccination serum samples were found to be significantly higher, but portions of the samples did not show seroconversion by neutralizing-antibody tests. Similar results were reported by Sakata *et al.* (1984). The current consideration of neutralizing antibody as the most reliable index of the immunity level to mumps virus infection needs, therefore, to be reconsidered. Moreover, neutralizing-antibody tests are labor intensive and time consuming.

In a recent study (Koskela, 1985) the humoral immunity induced by a live *Francisella tularensis* vaccine was measured by a complement-fixing enzyme-linked immunosorbent assay (CF-ELISA). For this purpose CF-ELISA with bacterial sonicate as the antigen, human AB serum as source of complement and alkaline–phosphatase-labeled anti-(human C3c) as the conjugate was developed for detecting antibodies to *F. tularensis*. With this method, the humoral responses after inoculation with a live attenuated tularemia vaccine appeared mainly 4 weeks after vaccination, the response time varying from 1 week to 2 months. CF antibodies were still detectable 18 months after vaccination. A corresponding method with peroxidase-labeled anti-(guinea-pig C3) as the conjugate and guinea-pig serum as the complement was also described for Chagas disease (Tandon *et al.*, 1979) and for brucellosis (Hinchliffe and Robertson, 1983).

B. VACCINES: QUALITY ASSESSMENT AND STANDARDIZATION

Precise methods of strain characterization are essential for understanding the epidemiology of infectious agents and for studies of vaccine efficiency. Enzyme immunoassays, particularly in conjunction with the use of monoclonal antibodies, are suitable for this purpose (Osterhaus *et al.*, 1983).

Monoclonal antibodies used in enzyme immunoassays allow the antigenic characterization of strains, both qualitatively and quantitatively. Osterhaus *et al.* (1983) developed a theoretical pattern-filling computer program for this purpose for which five criteria were taken into consideration: (i) the comparison method should not minimize observed differences; (ii) a large difference is more important than a number of small differences; (iii) a variable is necessary to indicate the degree of relationship of a strain to a reference strain; (iv) small differences are expected and should be accounted for; and (v) the method should be applicable to large numbers of samples.

Schemes have been designed to assess essential parameters of vaccines and to deliver the necessary information on the standardization and efficacy of vaccines in the field. Internal quality control is used to ensure the quality of vaccine batches and to monitor the reproducibility of an assay system, i.e., the batches produced should conform to certain design specifications. External quality assessment, on the other hand, questions particular results obtained against certain references and will tend to question the fundamental design. This can then result in an improvement in performance outside the laboratory (Hunter *et al.*, 1981).

Bias is often too high for immunoassays even though precision is acceptable. The suitability of standards for measuring bias is very important. Problems related to standards are twofold: miscalibration of the in-house standard with the primary standard, and stability of stored solutions of the standards (Hunter *et al.*, 1981). Bias (systematic errors introduced by methodological or reagent variability) can be distinguished from random errors caused by machine or operator failure and from inherent variability. The monitoring of this precision is of overwhelming importance (Ekins, 1979; Woodhead *et al.*, 1981). Even among expert laboratories differences in results can be excessive (Bozoky, 1963).

The International Union of Immunological Societies Standardization Committee has been working toward the standardization of immunological reagents (Batty, 1981). The benefits of standardization were clearly demonstrated by Rowe *et al.* (1970), and a number of standards are available from WHO. The primary use of standards is to ensure activity, e.g., seroconversion after vaccination (Kuno-Sakai, 1984). Standards also provide a basis for the establishment of minimal seroconversions required for protection and a meaningful tool for epidemiological investigations.

14

Future Prospects of Enzyme Immunodiagnosis

Enzyme immunodiagnosis has been expanding rapidly and has benefitted health care and basic research in numerous ways. This enthusiastic momentum has generated an incentive for further applications, and many industries have joined in the research and production of enzyme immunoassays. Despite the fact that this young discipline has had an extraordinary impact, many major medical centers have been slow in adopting this new technology (*J. Ch. Exp. Imm.*, Dec. 1981) on the grounds that many of the enzyme immunoassays have limited value and limited prospects of growth, or that current methodology works well and experience has been accumulated in their use. Others have predicted that the United States market for assays for ligands will be around $400,000,000,000 in 1985 (Aloisi and Hyun, 1983). This situation stems most probably from the dependence of immunodiagnosis laboratories on kits, reagents, and instrumentation for automation, whereas research laboratories have the capacity of developing assays and reagents for their particular needs. The recent rapid increase in momentum in the industry to provide kits based on enzyme immunoassays and the possibility of automation combined with interfaced computers for rapid and accurate data processing allow one to predict that this discrepancy will rapidly disappear.

Current EIA techniques can be modified to increase the sensitivity of the enzyme–substrate reaction, i.e., the K_m of the substrate–enzyme and the detectability of the product. Moreover, the immunologic and enzymatic activities are far from maximum for most of the enzyme conjugates offered commercially or which are prepared with the techniques generally used. The design of the assay will continue to evolve from the originally simple adaptations to obtain maximum specificity and detectability. The progress of enzyme immunodiagnosis will certainly not be limited to the perfection of current technology but will evolve to even greater sophistication. Much current equipment is of limited use and will be replaced during this innovation process. On the other hand, assays adapted to automation are expected to become important for screening and epidemiological studies.

Tests will be designed which are less specific than some of the current ones so that they can be better adapted to clinical situations. A clinician might be less interested in knowing which strain of virus causes a disease, as is increasingly possible with the advent of monoclonal antibodies, than in knowing whether a disease is caused by a virus. A common denominator for a virus infection, e.g., an interferon, could therefore be very helpful. For bacterial infection, products such as peptidoglycan or muramic acid are common and may be of use.

Another development which could be of great importance is the *in vivo* immunoassay whereby immunoreactants are used for ''homing'' to study the function of infectious agents and the specific products produced on infection of certain tissues.

It is important to recognize that assays have inherent limitations and will not be applicable, e.g., for every virus (McIntosh, 1983). Some viruses will persist at very low titers and will be difficult to detect with current antigen detection methods. Others do not circulate and will not be present in blood, and others will normally cause no or weak antibody responses, thus making nonsense of serodiagnosis. New approaches for such cases will be necessary.

Finally, the present research on the development of rapid homogeneous immunoassay systems without solid phase, which do not require the separation of free and bound label, is very promising and would find important application in medical diagnosis and biological sciences.

Acknowledgments

Permission for the reproduction of illustrations is gratefully acknowledged to Dr. J. P. Sloane [*Cancer* (1981) **47,** 1786], Dr. Christian Micheau Dr. Euan M. McMillan [*Am. J. Clin. Pathol.* (1981) **76,** 737], Drs. Seigo Yamamoto and Yoichi Minamishima [*J. Clin. Microbiol.* (1982) **15,** 1128], and Dr. D. M. Helfman [*Proc. Natl. Acad. Sci. U.S.A.* (1983) **80,** 31]. Our thanks are extended to Mr. G. B. Filosi for his assistance in the preparation of figures 4.1–4.11 and 5.1–5.5.

Bibliography

Aardoom, H. A., de Hoop, D., Iserief, C. O. A., Michel, M. F., and Stolz, E. (1982). *Br. J. Vener. Dis.* **58,** 359.

Aaskov, J. G., and Davies, C. E. A. (1979). *J. Immunol. Methods* **25,** 37.

Abramowitz, A., Livni, N., Morag, A., and Ravid, Z. (1982). *Arch. Pathol. Lab. Method* **106,** 115.

Ackerman, S. B., and Kelly, E. A. (1983). *J. Clin. Microbiol.* **17,** 410.

Adler-Storthz, K., Dreesman, G. R., Graham, D. Y., and Evans, D. G. (1985). *J. Immunoassay* **6,** 67.

Albert, W. H. W., Kleinhammer, G., Linke, R., Transwell, P., and Staehler, F. (1978). *In* "Enzyme-Linked Immunoassay of Hormones and Drugs" (B. S. Pal, ed.), p. 153. de Gruyter, Berlin.

Al-Laissi, E. N., and Mostratos, A. (1983). *J. Immunol. Methods* **58,** 127.

Aloisi, R. M., and Hyun, J. (1983). *In* "Immunodiagnostics" (R. M. Aloisi and J. Hyun, eds.), p. 9. Liss, New York.

Altschuh, D., and van Regenmortel, M. H. V. (1982). *J. Immunol. Methods* **50,** 99.

Alwine, J. C., Kemp, D. J., Parker, B. A., Reiser, J., Renart, J., Stark, G. R., and Wahl, G. M. (1979). *Methods Enzymol.* **68,** 220.

Ambroise-Thomas, P., and Desgeorges, P. T. (1980). *Bull. Soc. Pathol.* **73,** 89.

Ambroise-Thomas, P., Daveau, C., and Desgeorges, P. T. (1980). *Bull. Soc. Pathol.* **73,** 430.

Anders, R. F., Brown, G. F., and Edwards, A. (1983). *Proc. Natl. Acad. Sci. U.S.A.* **80,** 6652.

Anderson, P., and Moller, B. R. (1982). *Scand. J. Infect. Dis.* **14,** 19.

Ankerst, J., Christensen, P., Kjetlen, A., and Kronvall, G. (1974). *J. Infect. Dis.* **130,** 268.

Ansari, M. A., Pintozzi, R. L., Choi, Y. C., and Landove, R. F. (1981). *Am. J. Clin. Pathol.* **76,** 94.

Appleyard, S. T., Dunn, M. J., Dobowitz, V., Scott, M. L., Pittman, S. J., and Shotton, D. M. (1984). *Proc. Natl. Acad. Sci. U.S.A.* **81,** 776.

Araujo, F. G., and Remington, J. S. (1980). *J. Infect. Dis.* **141,** 144.

Araujo, F. G., Handman, E., and Remington, J. S. (1980). *Infect Immun.* **30,** 12.

Atkinson, D. E. (1966). *Annu. Rev. Biochem.* **35,** 85.

Austin, M. A., Monto, A. S., and Maassab, A. F. (1978). *Arch. Virol.* **58,** 345.

Avrameas, S., and Uriel, J. (1966). *Compt. Rend. Acad. Sci. Paris* **262,** 2543.

Avrameas, S., Ternynck, T., and Guesdon, J.-L. (1978). *Scand. J. Immunol.* **8,** 7.

Axen, R., Porath, J., and Ernbach, S. (1967). *Nature (London)* **214,** 1302.

Bacquet, C., and Twumasi, D. (1984). *Anal. Biochem.* **136,** 487.

Balachandran, N., Frame, B., Chernesky, M., Kraiselburd, Y., Kouri, D., Garcia, C., Lavery, C., and Rawls, W. E. (1982). *J. Clin. Microbiol.* **16,** 205.

Banatvala, J. E., Best, J. M., Kennedy, E. A., Smith, E. E., and Spence, M. E. (1967). *Br. Med. J.* **3,** 285.

Banks, P. M. (1979). *J. Histochem. Cytochem.* **27,** 1192.

Barbara, D. J., and Clark, M. F. (1982). *J. Gen. Virol.* **58,** 315.

Barlough, J. E., Jacobson, R. H., Downing, D. R., Marcella, K. L., Lynch, T. J., and Scott, F. W. (1983). *J. Clin. Microbiol.* **17,** 202.

Barnard, D. L., Johnson, F. B., and Richards, D. F. (1985). *J. Clin. Pathol.* **38,** 1158.

Batteiger, B., Newhall, W. J., and Jones, R. B. (1982). *J. Immunol. Methods* **55,** 297.

Batty, I. (1981). *In* "Immunoassays for the 80s" (A. Voller, A. Bartlett, and D. Bidwell, eds.), p. 193. University Park Press, Baltimore.

Baughn, R. E., Adams, C. B., and Musher, D. M. (1983). *Infect. Immun.* **42,** 585.

Beards, G. M., and Bryden, A. S. (1981). *J. Clin. Pathol.* **34,** 1388.

Beards, G. M., Campbell, A. D., Cottrell, N. R., Perris, J. S. M., Rees, N., Sanders, R. C., Shirley, J. A., Wood, H. C., and Flewett, T. H. (1984). *J. Clin. Microbiol.* **19,** 284.

Bellamy, K., Hodgson, J., Gardner, P. S., and Morgan-Capner, P. (1985). *J. Clin. Pathol.* **38,** 1150.

Benedict, A. A., and Yagaga, K. (1976). *In* "Comparative Immunology" (J. J. Marchalonis, ed.), p. 335. Wiley, New York.

Benjamin, D. R. (1974). *Appl. Microbiol.* **28,** 568.

Berkson, J. (1944). *J. Am. Statis. Assoc.* **41,** 70.

Berman, M., Alter, H. J., Ishak, K. G., Purcell, R. H., and Jones, E. A. (1979). *Ann. Intern. Med.* **91,** 1.

Bernstein, M. T., and Stewart, J. A. (1971). *Appl. Microbiol.* **21,** 680.

Berzofsky, J. A., and Schlechter, A. N. (1981). *Mol. Immunol.* **18,** 751.

Best, J. M., Palmer, S. J., Morgan-Capner, P., and Hodgson, J. (1984). *J. Virol. Methods* **8,** 99.

Beutin, L., Bode, L., Richter, T., Peltre, G., and Stephen, R. (1984). *J. Clin. Microbiol.* **19,** 371.

Beutler, E., and Blume, K. G. (1979). *Proc. Hematol.* **11,** 17.

Billings, P. B., and Hoch, S. O. (1983). *J. Immunol.* **131,** 347.

Billings, P. B., Hoch, S. O., White, P. J., Carson, D. A., and Vaughn, J. A. (1983). *Proc. Natl. Acad. Sci. U.S.A.* **80,** 7104.

Binns, M. M., Boursnell, M. E. G., Foulds, I. J., and Brown, T. D. K. (1985). *J. Virol. Methods* **11**(3), 265.

Bishop, R. F., Cipriani, E., Lund, J. S., Barnes, G. L., and Hosking, C. S. (1984). *J. Clin. Microbiol.* **19,** 447.

Bittner, M., Kupterer, P., and Morris, C. F. (1980). *Anal. Biochem.* **102,** 459.
Blondel, B., and Crainic, R. (1981). *Dev. Biol. Stand.* **47,** 335.
Bogulaski, R. C., Carrico, R. J., and Christner, J. E. (1979). US Patent 4, 134, 792.
Böhm, H.-J., Kopperschläger, G., Schulz, J., and Hoffmann, E. (1972). *J. Chromatogr.* **69,** 209.
Bohn, W. (1978). *J. Histochem. Cytochem.* **26,** 293.
Bohn, W. (1980). *J. Gen. Virol.* **46,** 439.
Bollum, F. J. (1975). *Proc. Natl. Acad. Sci. U.S.A.* **72,** 4119.
Bonfanti, C., Meurman, O., and Halonen, P. (1985). *J. Virol. Methods* **11,** 161.
Booth, J. C., Hannington, G., Aziz, T. A. G., and Stein, H. (1979). *J. Clin. Pathol.* **32,** 122.
Booth, J. C., Hannington, G., Bakir, T. M. F., Stein, H., Kangro, H. O., Griffiths, P. D., and Health, R. B. (1982). *J. Clin. Pathol.* **35,** 1345.
Bos, H. J., Schouten, W. J., Noordpool, H., Maklin, M., and Oostburg, B. F. J. (1980). *Am. J. Trop. Hyg.* **29,** 358.
Bosman, F. T., Cramer-Knynenburg, G., and van Bergen Henegouw, J. (1980). *Histochemistry* **67,** 243.
Boulard, C., and Lecroisey, A. (1982). *J. Immunol. Methods* **50,** 221.
Bowen, B., Steinberg, J., Laemmli, U. K., and Weintraub, H. (1980). *Nucl. Acids Res.* **8,** 1.
Bowman, J. D., and Aladjem, F. (1963). *J. Theor. Biol.* **4,** 242.
Bozoky, S. (1963). *Arthr. Rheum.* **6,** 641.
Brandon, D. L., Corse, J. W., Windle, J. J., and Layton, L. L. (1985). *J. Immunol. Methods* **78,** 87.
Brandt, C. D., Kim, H. W., Rodriguez, W. J., Thomas, L., Yolken, R. H., Arrobia, J. O., Kapikian, A. Z., Parrott, R. H., and Chanock, R. M. (1981). *J. Clin. Microbiol.* **13,** 976.
Brandt, C. D., Kim, H. W., Rodriguez, W. J., Arrobio, J. O., Jeffries, B. C., Stallings, E. P., Lewis, C., Miles, A. J., Chanock, R. M., Kapikian, A. Z., and Parrott, R. H. (1983). *J. Clin. Microbiol.* **18,** 171.
Brandtzaeg, P. (1976). *Clin. Exp. Immunol.* **25,** 50.
Brandtzaeg, P. (1982). *In* "Techniques in Immunochemistry" (G. R. Bullock and P. Petrusz, eds.), Vol. 1, p. 1. Academic Press, London.
Braun, L., Farmer, E. R., and Shak, K. V. (1983). *J. Med. Virol.* **12,** 187.
Brener, Z. (1982). *Bull. WHO* **60,** 463.
Briles, D. E., and Davie, J. M. (1980). *J. Exp. Med.* **152,** 151.
Broome, S., and Gilbert, W. (1978). *Proc. Natl. Acad. Sci. U.S.A.* **75,** 2746.
Bruce-Schwatt, L. J. (1979). *Trans. R. Soc. Trop. Med.* **73,** 605.
Bryan, F. L. (1969). *J. Milk Food Technol.* **32,** 381.
Buck, A. A., Andersen, R. I., and MacRae, A. A. (1978). *Tropenmed. Parasitol.* **29,** 137.
Bullock, G. R., and Petrusz, P. (1982). "Techniques in Immunochemistry," Vol. 1. Academic Press, New York.
Bundo, K., and Igarashi, A. (1985). *J. Virol. Methods* **11,** 15.
Burd, J. F., Carrico, R. J., Fetter, M. C., Buckler, R. T., Johnson, R. D., Bogulaski, R. C., and Christner, J. E. (1977). *Anal. Biochem.* **77,** 56.
Burke, D. S., and Nisalak, A. (1982). *J. Clin. Microbiol.* **15,** 353.
Burnette, W. N. (1981). *Anal. Biochem.* **112,** 195.
Burt, S. M., Carter, T. J. N., and Kricka, L. J. (1979). *J. Immunol. Methods* **31,** 231.
Burtin, P., Gold, P., Chu, T. M., Hammarström, S. G., Hansen, H. J., Johanssen, G. B.,

von Kleist, S., Mach, J-P., Neville, A. M., Shively, J. E., Stroebel, P., and Zamked, N. (1978). *Scand. J. Immunol.* **8,** 27.

Butler, J. E. (1981). *Methods Enzymol.* **73,** 482.

Butler, J. E., Cantarero, L. A., Swanson, P., and McGivern, P. L. (1980). *In* "Enzyme-Immunoassay" (E. G. Maggio, ed.), p. 197. CRC, Boca Raton, Florida.

Byrd, J. W., Heck, F. C., and Hidalgo, R. J. (1979). *Am. J. Vet. Res.* **40,** 891.

Calcott, M. A., and Müller-Eberhard, H. J. (1971). *Biochemistry* **11,** 3443.

Cantarero, L. A., Butler, J. E., and Osborne, J. W. (1980). *Anal. Biochem.* **105,** 375.

Cappel, R. L., Cowman, A. F., Lingelback, K. R., Brown, G. V., Saint, R. B., Kemp, D. J., and Anders, R. F. (1983). *Nature (London)* **306,** 751.

Capra, J. D., Tung, A. S., and Nisonoff, A. (1975). *J. Immunol.* **114,** 1548.

Capron, A., Yarzabal, L., Vernes, A., and Truit, J. (1980). *Pathol. Biol.* **18,** 357.

Carbone, A., Micheau, C., Caillaud, J-M., and Carlu, C. (1981). *Cancer* **47,** 2862.

Carlson, J., and Eriksson, S. (1980). *Acta Med. Scand.* **207,** 79.

Carlson, J., Eriksson, S., and Hägerstrand, I. (1981). *J. Clin. Pathol.* **34,** 1020.

Carlsson, B., Ahlstedt, S., Hanson, L. A., Liden-Janson, G., Lindblad, B. S., and Sultana, R. (1976). *Acta Paediatr. Scand.* **65,** 417.

Carruthers, M. M., Jenkins, K. E., Kabat, W. J., and Buranosky, T. (1984). *J. Clin. Microbiol.* **19,** 552.

Catanzaro, P. J., Brandt, W. E., Hogrefe, W. R., and Russell, P. K. (1974). *Infect. Immun.* **10,** 381.

Cesbron, J. Y., Capron, A., Ovlaque, G., and Santono, F. (1985). *J. Immunol. Methods* **83,** 151.

Chang, H. C., Takashima, I., Arikawa, J., and Hashimoto, N. (1984). *J. Immunol. Methods* **72,** 401.

Chang, J. J., Crowl, C. P., and Schneider, R. S. (1975). *Clin. Chem.* **21,** 967.

Chanock, S. J., Wenske, E. A., and Fields, B. N. (1983). *J. Infect. Dis.* **148,** 49.

Chao, R. K., Fisher, M., Schwarsmen, J. D., and McIntosh, K. (1979). *J. Infect. Dis.* **139,** 483.

Charpentier, G., Garzon, S., and Kurstak, E. (1982). *Ann. Virol. Inst. Pasteur,* **133E,** 223.

Chessum, B. S., and Denmark, J. R. (1978). *Lancet* **1,** 161.

Christie, K. E., and Haukenes, G. (1983). *J. Med. Virol.* **12,** 267.

Claftin, J. L., and Williams, K. (1978). *Curr. Top. Microbiol. Immunol.* **81,** 107.

Clark, W. A. (1983). *Bio-Radiations (Bio-Rad Labs)* **45,** 1.

Cobb, M. E., Buckley, N., Hu, M. W., Miller, J. G., Singh, P., and Schneider, R. S. (1977). *Clin. Chem.* **23,** 1161.

Cohen, J. (1979). *J. Gen. Virol.* **36,** 395.

Coleman, M. S., Greenwood, M. F., and Hutton, J. J. (1976). *Cancer Res.* **36,** 120.

Collins, J. K., and Kelly, M. T. (1983). *J. Clin. Microbiol.* **17,** 1005.

Conradie, J. D., Govender, M., and Visser, L. (1983). *J. Immunol. Methods* **59,** 289.

Coons, A. H., Snyder, J. E., Cheever, F. S., and Murray, E. J. (1950). *J. Exp. Med.* **91,** 31.

Cooperative Study (1975). *Arch. Dermatol.* **111,** 371.

Cost, K. M., West, C. S., Brinson, D., and Polk, H. C. (1985). *J. Immunoassay* **6,** 23.

Coulson, B. S., and Holmes, I. H. (1984). *J. Virol. Methods* **8,** 165.

Cox, J. C., Moloney, M. B., Herrington, R. W., Hampson, A. W., and Hurrell, J. G. R. (1984). *J. Virol. Methods* **8,** 137.

Cradock-Watson, J. E., Ridehalgh, M. K. S., and Bourne, M. S. (1979). *J. Hyg.* **28,** 319.

Crane, M. S. J., and Dvorak, J. A. (1980). *Science* **208,** 194.

Cremer, N. E., Cossen, C. K., Hanson, C. V., and Shell, G. R. (1982). *J. Clin. Microbiol.* **13,** 226.

Crook, N. E., and Payne, C. C. (1980). *J. Gen. Virol.* **46,** 29.

Crum, J. W., Hanchalay, S., and Eamsila, C. (1980). *J. Clin. Microbiol.* **11,** 584.

Cukor, G., and Blacklow, N. R. (1984). *Microbiol. Rev.* **48,** 157.

Cukor, G., Perron, D. M., Hudson, R., and Blacklow, N. R. (1984). *J. Clin. Microbiol.* **19,** 888.

Culling, C. F. A. (1974). "Handbook of Histopathological and Histochemical Techniques," 3rd ed. Butterworths, London.

Cunningham, M. W., and Russell, S. M. (1983). *Infect. Immun.* **42,** 531.

Curtis, E. G., and Patel, J. A. (1978). *CRC Crit. Rev. Clin. Lab. Sci.* **9,** 303.

Danhof, M., and Breimer, D. D. (1978). *Clin. Pharmak.* **3,** 39.

Danielsson, B., Battiasson, B., and Mosback, K. (1983). *Pure Appl. Chem.* **51,** 1443.

Dasch, G. A. (1981). *J. Clin. Microbiol.* **14,** 333.

Dasch, G. A., Halle, S., and Bourgeoi, A. L. (1979). *J. Clin. Microbiol.* **9,** 38.

Da Silva, L. R., Locke, M., Dayal, R., and Perrin, L. A. (1983). *Bull. WHO* **61,** 105.

Davies, M. E., Barrett, A. J., and Hembry, R. M. (1978). *J. Immunol. Methods* **21,** 305.

DeBlas, A. L., and Cherwinski, H. M. (1983). *Anal. Biochem.* **133,** 214.

De Clercq, E., Deschamps, J., Verhelst, G., Walker, R. T., Jones, A. S., Torrence, P. F., and Shugar, D. (1980). *J. Infect. Dis.* **141,** 563.

Deinhardt, F., and Gust, I. D. (1982). *Bull. WHO* **60,** 661.

de Jong, J. C., Wigand, R., Kidd, A. H., Wadell, G., Kapsenberg, J. G., Muserie, C. J., Wermenbol, A. G., and Firtzlaff, R. G. (1983). *J. Med. Virol.* **11,** 215.

de la Llosa, P., El Abed, A., and Roy, M. (1980). *Can. J. Biochem.* **58,** 745.

De Lean, A., Munson, P. J., and Rodbard, D. (1978). *Am. J. Physiol.* **235**(2), E97.

DeLellis, R. A., Sternberger, L. A., Mann, R. B., Banks, P. M., and Nakane, P. K. (1979). *Am. J. Clin. Pathol.* **71**(5), 483.

Denk, H., Tappeiner, C., Eckerstorger, R., and Holzner, J. H. (1973). *Digestion* **9,** 106.

de Porceri-Morton, C., Chang, J., Specker, M., and Bastiani, R. (1980). "Clinical Study 50." Syva Co. Palo Alto, Ca.

de Savigny, D. H. (1975). *J. Parasitol.* **61,** 281.

de Savigny, D. H., and Voller, A. J. (1980). *J. Immunoassay* **1,** 105.

de Savigny, D. H., Voller, A., and Woodruff, A. W. (1979). *J. Clin. Pathol.* **32,** 543.

Des Moutis, I., Ouaissi, A., Grzych, J. M., Yarzabal, L., Hague, A., and Capron, A. (1983). *Am. J. Trop. Med. Hyg.* **32,** 533.

Dienstag, J. L., Feinstone, S. M., Purcell, R. H., Wang, D. C., Alter, H. J., and Holland, P. V. (1977). *Lancet* **1,** 560.

Dienstag, J. L., Krotoski, W. A., Howard, W. A., Purcell, R. H., Neva, F. A., Galambos, J. T., and Glew, R. H. (1981). *J. Infect. Dis.* **143,** 200.

Dietzler, D. N., Weidman, N., and Tieber, V. L. (1980). *Clin. Chim. Acta* **101,** 163.

Dietzler, D. N., Leckie, M. P., Hoelting, C. R., Porter, S. E., Smith, C. H., and Tieber, V. L. (1983). *Clin. Chem. Acta* **127,** 239.

Diwan, A. R., Coker-Vann, M., Brown, P., Subianto, D. B., Yolken, R., Desowitz, R., Escobar, A., Gibbs, C. J., Jr., and Gajdusek, D. C. (1982). *Am. J. Trop. Med. Hyg.* **31,** 364.

Dobbins Place, J., and Schroeder, R. (1980). *J. Immunol. Methods* **48,** 251.

Donehower, R. C., Hande, K. R., Drake, J. C., and Chabner, B. A. (1979). *Clin. Pharmacol. Ther.* **36,** 63.

Dormeyer, H. H., Arnold, W., Kryger, P., Nielsen, J. O., and Meyer, K. H. (1981). *Klin. Wochenschr.* **59,** 675.

Dörries, R., and ter Meulen, V. (1983). *J. Gen. Virol.* **64,** 159.

Dougherty, R. M., DiStefano, H. S., and Marucci, A. A. (1974). *In* "Viral Immunodiagnosis" (E. Kurstak and R. Morisset, eds), p. 89. Academic Press, New York.

Dreesman, G. R., Hollinger, F. B., and Melnick, J. L (1975). *Am. J. Med. Sci.* **270,** 123.

Drew, D. L., Maki, D. G., and Manning, D. D. (1979). *J. Clin. Microbiol.* **10,** 339.

Duermeyer, W., and van der Veen, J. (1978). *Lancet* **2,** 684.

Duermeyer, W., Wielaard, F., and van der Veen, J. (1979). *J. Med. Virol.* **4,** 25.

Duermeyer, W., Stute, R., and Hellings, J. A. (1983). *J. Med. Virol.* **11,** 11.

Duhamel, R. C., Schur, P. H., Brendel, K., and Meesan, E. (1979). *J. Immunol. Methods* **31,** 211.

Dyer, J. R. (1956). *Methods Biochem. Anal.* **3,** 111.

Eberle, R., and Mou, S. W. (1983). *J. Infect. Dis.* **148,** 436.

Ehrlich, P. H., Moyle, W. R., Moustafa, Z. A., and Canfield, R. E. (1982). *J. Immunol.* **128,** 2709.

Eisen, H. N. (1980). "Immunology: An Introduction to Molecular and Cellular Principles of the Immune Responses," 2nd ed. Harper and Row, Hogerstown, Md.

Ekins, R. P. (1979). *In* "Radioimmunology" (C. A. Bizollen, ed.), p. 239. Elsevier Biomedical Press, Amsterdam.

Endres, D. B., Painter, K., and Niswender, G. D. (1978). *Clin. Chem.* **24,** 460.

Engvall, E. (1977). *Med. Biol.* **55,** 193.

Engvall, E., and Ljungstrom, I. (1975). *Acta Pathol. Microbiol. Scand. C* **83,** 231.

Erickson, P. F., Minier, L. N., and Lasher, R. S. (1982). *J. Immunol. Methods* **51,** 241.

Esen, A., Conroy, J. M., and Wang. S.-Z. (1983). *Anal. Biochem.* **132,** 462.

Espinoza, B., Flisser, A., Plancarle, A., and Lairalde, L. (1982). *In* "Cysticercosis: Present State of Knowledge and Perspectives" (A. Flisser, J. Willms, J. P. Laclette, C. Lauralde, C. Ridaura, and F. Beltram, eds.), p. 163. Academic Press, New York.

Evans, W. E., Pratt, C. B., Taylor, R. H., Barker, L. F., and Crem, W. R. (1979). *Cancer Chemother. Parmocol.* **3,** 161.

Ey, P. L., Prowse, S. J., and Jenkin, C. R. (1978). *Immunochemistry* **15,** 429.

Farr, A. G., and Nakane, P. K. (1981). *J. Immunol. Methods* **47,** 129.

Feinstone, S. M., Kapikian, A. Z., and Purcell, R. H. (1973). *Science* **182,** 1026.

Felgner, P. (1978). *Zbl. Bakt. Hyg. 1, Abt. Orig. A* **242,** 100.

Ferrua, B., Marolini, R., and Masseyelf, R. (1979). *J. Immunol. Methods* **25,** 49.

Fey, H. (1981). *J. Immunol. Methods* **47,** 109.

Fey, H., Pfister, H., and Ruegg, O. (1984). *J. Clin. Microbiol.* **19,** 34.

Finn, M. P., Ohlin, A., and Schachter, J. (1983). *J. Clin. Microbiol.* **17,** 848.

Flewett, T. H., and Babiuk, L. A. (1984). *In* "Control of Virus Diseases" (E. Kurstak and R. G. Marusyk, eds.), pp. 57–65. Dekker, New York.

Flisser, A., Perez-Montfort, R., and Larralde, C. (1979). *Bull. WHO* **57,** 839.

Forghani, B., and Schmidt, N. J. (1979). *J. Clin. Microbiol.* **9,** 657.

Forghani, B., Schmidt, N. J., and Lennette, E. H. (1973). *Intervirology* **1,** 48.

Forghani, B., Schmidt, N. J., and Lennette, E. H. (1974). *Appl. Microbiol.* **28,** 661.

Forghani, B., Schmidt, N. J., and Dennis, J. (1978). *J. Clin. Microbiol.* **8,** 545.

Forghani, B., Dennis, J., and Schmidt, N. J. (1980). *J. Clin. Microbiol.* **12,** 704.

Frame, B., Mahony, J. B., Balachandran, N., Rawls, W. E., and Chernesky, M. A. (1984). *J. Clin. Microbiol.* **20,** 162.

Freeman, R. R., and Holder, A. A. (1983). *J. Exp. Med.* **158,** 1647.
Friedman, E., Thor, A., HoranHand, P., and Schlom, I. (1985). *Cancer Res.* **45,** 5648.
Friguet, B., Djavadi Ohaniance, L., Pages, J., Bussard, A., and Goldberg, M. (1983). *J. Immunol. Methods* **60,** 351.
Fujiwara, K., Ono, S., Fujinaka, H., and Kitagawa, T. (1984). *J. Immunol. Methods* **72,** 109.
Gallati, H., and Brodbeck, H. (1982). *J. Clin. Chem. Clin. Biochem.* **20,** 221.
Gamble, H. R., and Graham, C. E. (1984). *Vet. Immunol.* **6,** 379.
Geiseler, P. J., Morris, R. D., and Harris, B. (1985). *In* "Advances in Sexually Transmitted Diseases" (R. Morisset and E. Kurstak, eds.), pp. 87–94. VNU Science Press, Holland.
Geoghegan, W. D., Struve, M. F., and Jordan, R. E. (1983). *J. Immunol. Methods* **60,** 61.
Gerety, R. J., ed. (1981). "Non-A, Non-B Hepatitis." Academic Press, New York.
Gerhard, J., and Webster, R. G. (1978). *J. Exp. Med.* **148,** 383.
Gerhard, W., Croce, C. M., Lopes, D., and Koprowski, H. (1978). *Proc. Natl. Acad. Sci. U.S.A.* **75,** 1510.
Gerna, G., and Chambers, R. W. (1977a). *Intervirology* **8,** 257.
Gerna, G., and Chambers, R. W. (1977b). *J. Med. Microbiol.* **10,** 309.
Gershoni, J. M., and Palade, G. E. (1983). *Anal. Biochem.* **131,** 1.
Ghose, L. H., Schnagl, R. D., and Holmes, I. H. (1978). *J. Clin. Microbiol.* **8,** 268.
Giallongo, A., Kochoumian, L., and King, T. P. (1982). *J. Immunol. Methods* **52,** 379.
Girard, R., and Goichot, J. (1981). *Ann. Immunol.* **132C,** 211.
Gissmann, L., and zur Hausen, H. (1980). *Int. J. Cancer* **25,** 605.
Gissmann, L., Diehl, V., Schultz-Loulen, H. J., and zur Hausen, H. (1982). *J. Virol.* **44,** 393.
Glass, W. F., Briggs, R. C., and Hnilica, L. S. (1981). *Science* **211,** 70.
Glickman, L., Schantz, P., Dambrocke, R., and Cypress, R. (1979). *Am. J. Trop. Med. Hyg.* **27,** 492.
Glynn, A. A., and Ison, C. (1981). *In* "Immunoassays for the 80s" (A. Voller, H. Bartlett, and D. Bidwell, eds.), p. 431. University Park Press, Baltimore, Md.
Gold, D., Gold, M., and Freedman, S. O. (1968). *Cancer* **23,** 1331.
Goldenberg, D. M., Sharkey, R. M., and Primus, F. J. (1978). *Cancer* **42,** 1546.
Goldman, R. C., White, D., Orshov, F., Orshov, I., Rick, P. D., Lewis, M. S., Battachaijel, A. K., and Leive, L. (1982). *J. Bacteriol.* **151,** 1210.
Goldwater, P. N., Webster, M., and Banatvala, J. E. (1982). *J. Virol. Methods* **4,** 9.
Gonwa, T. A., Peterlin, P. M., and Stobo, J. D. (1983). *Adv. Immunol.* **34,** 71.
Goodman, H. M. (1979). *Nature (London)* **281,** 351.
Gordon, J., Towbin, H., and Rosenthal, M. (1982). *J. Rheumatol.* **9,** 247.
Gordon, J., and Rosenthal, M. (1984). *J. Rheumatol.* **11,** 231.
Gordon, L. K. (1982). *J. Immunol. Methods* **44,** 241.
Gottlieb, M. S., Schroff, R., Schanker, H. M., Weissman, J. D., Fan, P. T., Woll, R. A., and Saxon, A. (1981). *N. Engl. J. Med.* **305,** 1425.
Greenberg, H. B., Kalica, A. R., Wyatt, R. G., Jones, R. W., Kapikian, A. Z., and Chanock, R. M. (1981a). *Proc. Natl. Acad. Sci. U.S.A.* **78,** 420.
Greenberg, H. B., Valdesucy, J. R., Kalica, A. R., Wyatt, R. G., McAuliffe, V. J., Kapikian, A. Z., and Chanock, R. M. (1981b). *J. Virol.* **37,** 994.
Greenberg, H. B., Wyatt, R. G., Kalica, A. R., Yolken, R. H., Black, R., Kapikian, A. Z., and Chanock, R. M. (1981c). *Perspect. Virol.* **11,** 163.
Greenwood, B. M., Whittle, H. C., and Dominic-Rajkovic, O. (1971). *Lancet* **2,** 519.

Greenwood, H. M., and Chandler, J. (1980). *In* "Centrifugal Analysis in Clinical Chemistry" (C. P. Price and K. Spencer, eds.), p. 125. Prayer, Eastbourne.

Gregory, J. F., Manley, D. M., and Kirk, J. R. (1981). *J. Agric. Food Chem.* **29,** 921.

Gripenberg, M. F., Wafin, F., Isomali, H., and Linder, E. (1979). *J. Immunol. Methods* **31,** 109.

Grom, J., and Bernard, S. (1985). *J. Virol. Methods* **10,** 135.

Guesdon, J.-L., and Avrameas, S. (1980). *J. Immunol. Methods* **39,** 1.

Guesdon, J.-L., Ternynck, T., and Avrameas, S. (1979). *J. Histochem. Cytochem.* **27,** 1131.

Guimaraes, M. C. S., Celeste, B. J., de Castilho, E. A., Minco, J. R., and Diner, J. M. P. (1981). *Am. J. Trop. Med. Hyg.* **30,** 942.

Gupta, J. D., Petersen, V., Stout, M., and Murphy, A. M. (1971). *J. Clin. Pathol.* **24,** 547.

Gupta, R. C. (1982). *Clin. Exp. Immunol.* **49,** 543.

Gupta, S. K., Guesdon, J. L., Avrameas, S., and Talwar, G. P. (1985). *J. Immunol. Methods* **83,** 159.

Gust, I. D., Dienstag, J. L., Purcell, R. H., and Lucas, C. R. (1977). *Br. Med. J.* **1,** 193.

Haan, P. de., Bakker, W., Boorsma, D. M., Heudekamp, F., and Kalsbeek, G. L. (1978). *Dermatologica* **156,** 303.

Haan, P. de., Boorsma, D. M., and Kalsbeek, G. L. (1979). *Allergy* **34,** 111.

Haggarty, A., Legler, C., Krantz, M. J., and Fuks, A. (1986). *Cancer Res.* **46,** 300.

Halbert, S. P.. Bastomsky, C. H., and Anken, M. (1983). *Clin. Chem. Acta* **127,** 69.

Hall, W. W., and Choppin, W. W. (1981). *N. Engl. J. Med.* **304,** 1152.

Halverson, C. A., Folini, B., Taylor, C. R., and Parker, J. W. (1981). *Am. J. Pathol.* **105,** 241.

Hamaguchi, Y., Yoshitake, S., Ishikawa, E., Endo, Y., and Ohtaki, S. (1979). *J. Biochem.* **85,** 1289.

Hammond, G. W., Smith, S. J., and Noble, G. R. (1980). *J. Infect. Dis.* **141,** 644.

Hammond, G. W., Ahluwalia, G. S., Baker, F. G., Horsman, G., and Hazelton, P. R. (1982). *J. Clin. Microbiol.* **16,** 53.

Hanahan, D. (1983). *J. Mol. Biol.* **166,** 557.

Hanahan, D., and Meselson, M. (1980). *Gene* **10,** 63.

Hanff, P. A., Fehniger, T. E., Miller, J. N., and Loveti, M. A. (1983). *J. Immunol.* **129,** 1287.

Harboe, N. M. G., and Ingild, A. (1983). *Scand. J. Immunol.* **17**(10), 345.

Harcourt, G. C., Best, J. M., and Banatvala, J. E. (1980). *J. Infect. Dis.* **142,** 145.

Hardin, J. A., and Thomas, J. O. (1984). *Proc. Natl. Acad. Sci. U.S.A.* **80,** 7410.

Harmon, M. W., and Pawlik, K. M. (1982). *J. Clin. Microbiol.* **15,** 5.

Harrowe, D. J., and Taylor, C. R. (1981). *J. Surg. Oncol.* **16,** 1.

Hasida, S., Imagawa, M., Eshikawa, E., and Freytag, J. W. (1985). *J. Immunoassay* **6,** 111.

Hawkes, R., Niday, E., and Gordon, J. (1982). *Anal. Biochem.* **119,** 142.

Hecht, T., Forman, S. J., Bross, K. J., Schmidt, G. M., and Blume, K. (1981). *New Engl. J. Med.* **304,** 848.

Heck, F. C., Williams, J. D., and Pruett, J. (1980). *J. Clin. Microbiol.* **11,** 398.

Heinz, F. X., Roggendorf, M., Hofmann, H., Junz, C., and Deinhardt, F. (1981). *J. Clin. Microbiol.* **11,** 581.

Helfman, D. M., Feramisco, J. R., Fiddes, J. C., Thomas, G. P., and Hughes, S. H. (1983). *Proc. Natl. Acad. Sci. U.S.A.* **80,** 31.

Hennessy, K., and Kieff, E. (1983). *Proc. Natl. Acad. Sci. U.S.A.* **80,** 5665.

Herbert, W. J. (1973). *In* "Handbook of Experimental Immunology" (D. M. Weir, ed.), 2nd ed., Appendix 2 and 3. Blackwell Scientific Publications, Oxford.

Hermann, J. E., Hendry, R. M., and Collins, M. F. (1979). *J. Clin. Microbiol.* **10,** 210.
Herscher, J. L., Siegel, R. J., Said, J. W., Ewalds, G. M., Moran, M. M., and Fishbein, M. C. (1984). *Am. J. Clin. Pathol.* **81,** 198.
Heyderman, E. (1979). *J. Clin. Pathol.* **32**(10), 971.
Heyderman, E., and Neville, A. M. (1977). *J. Clin. Pathol.* **30,** 138.
Higashi, G. I. (1984). *Diagn. Immunol.* **2,** 2.
Hillyer, G. V., and Gomez de Rios, I. (1979). *Am. J. Trop. Med. Hyg.* **28,** 237.
Hillyer, G. V., and Pelley, R. P. (1980). *Am. J. Trop. Med.* **29,** 582.
Hinchliffe, P. M., and Robertson, L. (1983). *J. Clin. Pathol.* **36,** 100, 22.
Ho, A. D., Helmstädter, V., and Hunstein, W. (1982). *Klin. Wochenschrift* **60,** 451.
Hofmann, H., Frisch-Niggemeyer, W., Heinz, F. (1979). *J. Gen. Virol.* **42,** 505.
Hogg, R. J., and Davidson, G. P. (1982). *Aush. J. Paediatr.* **18,** 184.
Hohmann, A. W., and Faulkner, P. (1983). *Virology* **125,** 432.
Holborow, E. J., Weir, D. M., and Johnson, G. D. (1957). *Br. Med. J.* **2,** 732.
Hollinger, F. B., and Dienstag, J. L. (1980). *In* "Manual of Clinical Microbiology" (E. H. Lennette, A. Balows, W. J. Hausler, Jr., and J. P. Truant, eds.), 3rd ed. p. 899. American Society of Microbiology, Washington, D.C.
Holmes, K. K. (1981). *J. Am. Med. Assoc.* **245,** 1718.
Holmes, R. K., and Sanford, J. P. (1974). *J. Infect. Dis.* **129,** 519.
Holmgren, J., and Svennerholm, A. M. (1973). *Infect. Immun.* **7,** 759.
Hornsleth, A., Grauballe, P. C., Friis, B., Genner, J., and Pedersen, I. R. (1981). *J. Clin. Microbiol.* **14,** 510.
Hornsleth, A., Friis, B., Andersen, P., and Brenøe, E. (1982). *J. Med. Virol.* **10,** 273.
Hornsleth, A., Friis, B., Grauballe, P. C., and Krasilnikof, P. A. (1984). *J. Med. Virol.* **13,** 149.
Hovi, T., Vaisanon, V., Ukkonen, P., and von Bonsdorf, D. H. (1982). *J. Virol. Methods* **5,** 45.
Howe, J., Wyse, B. W., and Hansen, R. G. (1979). *Fed. Proc.* **38,** 452.
Hsu, S., Raine, L., and Fanger, M. (1981). *J. Histochem. Cytochem.* **29,** 577.
Huang, S. N. (1975). *Lab. Invest.* **3,** 88.
Hughes, W. T. (1978). *Johns Hopkins Med. J.* **143,** 184.
Huitric, E., Laumonier, R., Burtin, D., von Kleist, S., and Chavanel, G. (1976). *Lab. Invest.* **34,** 97.
Hunter, W. M., McKenzie, I., and Bacon, R. R. A. (1981). *In* "Immunoassays for the 80s" (A. Voller, A. Bartlett, and D. Bidwell, eds.), p. 155. University Park Press, Baltimore.
Inouye, S., Matsuno, S., and Yamaguchi, H. (1984). *J. Clin. Microbiol.* **19,** 259.
Isaac, M., and Payne, R. A. (1982). *J. Med. Virol.* **10,** 55.
Isaacson, P., and Judd, M. A. (1977). *Gut* **18,** 786.
Ishikawa, E., Hamaguchi, Y., and Imayawa, M. (1980). *J. Immunoassay* **1,** 385.
Ishikawa, E., Imagawa, M., Hashida, S., Yoshitake, S., Hamaguchi, Y., and Ueno, T. (1983). *J. Immunoassay* **4,** 209.
Isobe, Y., Chen, S.-T., Nakane, P. K., and Brown, W. R. (1972). *Acta Histochem. Cytochem* **10,** 161.
Ito, J. I., Wunderlich, A. C., Lyons, J., Davis, C. E., Guiney, D. G., and Braude, A. I. (1980). *J. Infect. Dis.* **142,** 532.
Ito, M., and Barron, A. L. (1974). *Proc. Soc. Exp. Biol. Med.* **146,** 41.
Ionescu-Matia, I., Sanchez, Y., Fields, H. A., and Dreesman, G. R. (1983). *J. Virol. Methods* **6,** 41.

Izutsu, A., Leuz, D., Araps, C., Singh, P., Joklitsch, A., and Kabakoff, D. S. (1979). *Clin. Chem.* **25,** 1093.
Jackson, S. G., Yip-Chuck, D. A., and Brodsky, M. H. (1985). *J. Immunol. Methods* **83,** 141.
Jäger, G. (1981). *Blut* **42,** 259.
Jalkanen, M., and Jalkanen, S. (1983). *J. Clin. Lab. Immunol.* **10,** 225.
Jambazian, A., and Holper, J. C. (1972). *Proc. Soc. Exp. Biol. Med.* **140,** 560.
Janossy, G., Hoffbrand, A. V., Greaves, M. F., Ganesharaguru, K., Pain, C., Bradstock, K., Prentice, H. G., and Kay, H. E. M. (1980). *Br. J. Haematol.* **44,** 221.
Jeanssen, S. (1974). *Proc. Soc. Exp. Biol. Med.* **147,** 788.
Jensenius, J. C., Andersen, I., Hau, J., Crone, M., and Koch, C. (1981). *J. Immunol. Methods* **46,** 63.
Jessen, K. R. (1983). *In* "Immunochemistry: Practical Applications in Pathology and Biology" (J. M. Polak and S. Van Noorden, eds.), p. 169. Wright P.S.G., Bristol.
Johnson, R. B., Jr., and Libby, R. (1980). *J. Clin. Microbiol.* **12,** 451.
Jones, R. B., Ardery, B. R., Hui, S. L., and Cleary, R. E. (1982). *Fertil. Stril.* **38,** 553.
Jones, R. B., Bruins, S. C., and Newhall, V. W. J. (1983). *J. Clin. Microbiol.* **17,** 466.
Juhl, B. R., Nørgaard, K., and Bjerrum, O. J. (1984). *J. Histochem. Cytochem.* **32,** 935.
Julkunen, I. (1984). *J. Med. Virol.* **14,** 177.
Julkunen, I., Pyhala, R., and Hovi, T. (1985). *J. Virol. Methods* **10,** 75.
Kabakoff, D. S., Leug, D., and Sigh, P. (1978). *Clin. Chem.* **24,** 1055.
Kabakoff, D. S., and Greenwood, L. M. (1982). *Rec. Adv. Clin. Biochem.* **2,** 1.
Kachani, F., and Gocke, D. J. (1973). *J. Immunol.* **111,** 1564.
Kalica, A. R., Greenberg, B., Wyatt, R. G., Flores, J., Sereno, M. M., Kapikian, A. Z., and Chanock, R. M. (1981). *Virology* **112,** 385.
Kangro, H. O., Booth, J. C., Bakir, T. M. F., Tryhorn, Y., and Sutherland, S. (1984). *J. Med. Virol.* **14,** 73.
Kanno, T., Suzuki, H., Katsushima, N., Imai, A., Tazawa, F., Kutsuzawa, T., Kitaoka, S., Sakamoto, M., Yasaki, N., and Ishida, N. (1981). *J. Infect. Dis.* **147,** 125.
Kaper, J. W., Hagenaars, A. M., and Notermans, S. (1980). *J. Food Safety* **2,** 35.
Kapikian, A. Z., Wyall, R. G., Dolin, R., Thornhill, T. S., Kalica, A. R., and Chanock, R. M. (1972). *J. Virol.* **10,** 1075.
Kaplan, D. A., Naumonski, L., and Collier, R. J. (1981). *Gene* **13,** 211.
Kaplan, S. L., Mason, E. O., Johnson, G., Broughton, R. A., Hurley, D., and Parke, J. C. (1983). *J. Clin. Microbiol.* **18,** 1201.
Karush, F. (1962). *Adv. Immunol.* **2,** 1.
Katze, G., and Crowell, R. L. (1980). *J. Gen. Virol.* **50,** 357.
Kelsoe, G. H., and Weller, T. H. (1978). *Proc. Natl. Acad. Sci. U.S.A.* **75,** 5715.
Kemp, D. J., Cappel, R. I., Cowman, A. F., Saint, R. B., Brown, G. V., and Anders, R. F. (1983). *Proc. Natl. Acad. Sci. U.S.A.* **80,** 3787.
Kennedy, R. C., Melnick, J. L., and Dreesman, G. R. (1984). *Science* **23,** 930.
Kenny, G. E., and Dunsmoor, C. L. (1983). *J. Clin. Microbiol.* **17,** 655.
Khan, M. W., Gallagher, M., Bucher, D., Cerini, C. P., and Kilbourne, E. D. (1982a). *J. Clin. Microbiol.* **16,** 115.
Khan, M. W., Bucher, D. J., Koul, A. K., Smith, E. D., and Kilbourne, E. D. (1982b). *J. Clin. Microbiol.* **16,** 813.
Killington, R. A., Newhook, L., Balachandran, N., Rawls, W. E., and Bachatti, S. (1981). *J. Virol. Methods* **2,** 223.

Klein, J. (1979). *Science* **203,** 516.
Klein, J. (1982). "Immunology. The Science of Self-Nonself Discrimination." Wiley. New York.
Klinman, N., and Press, J. (1975). *Transplant. Rev.* **24,** 41.
Klipstein, F. A., Engert, R. F., Houghten, R. A., and Rowe, B. (1984). *J. Clin. Microbiol.* **19**(6), 798.
Klotz, I. M., and Hunston, D. L. (1975). *J. Biol. Chem.* **250,** 3001.
Knell, J. D., Summers, M. D., and Smith, G. E. (1983). *Virology,* **125,** 381.
Knecht, D. A., and Dimond, R. L. (1984). *Anal. Biochem.* **136,** 180.
Koenig, R., and Paul, H. L. (1982). *J. Virol. Methods* **5,** 113.
Köhler, G. (1976). *Eur. J. Immunol.* **6,** 340.
Konishi, E., and Yamaoka, M. (1983). *J. Virol. Methods* **7,** 21é.
Konno, T., Suzuki, H., Imai, A., and Ishida, N. (1977). *J. Infect. Dis.* **135,** 259.
Koper, J. W., Hagenaars, A. M., and Notermans, S. (1980) *J. Food Safety* **2,** 35.
Koprowski, H., Herlyn, D., Lubeck, M., De Freitas, E., and Sears, H. F. (1984). *Proc. Natl. Acad. Sci. U.S.A.* **81,** 216.
Koskela, M., and Leinonen, M. (1981). *J. Clin. Pathol.* **34,** 93.
Koskela, P. (1985). *Vaccine* **3**(5) 389.
Koup, J. R., and Brodsky, B. (1978). *Am. Res. Resp. Dis.* **117,** 1125.
Kraaijeveld, C. A., Reed, S. E., and Macnaughton, M. R. (1980). *J. Clin. Microbiol.* **12,** 493.
Krause, P. J., Hyams, J. S., Middleton, P. J., Hersen, V. C., and Flores, J. (1983). *J. Pediatr.* **10,** 259.
Krech, V., and Jung, M. (1971). *Arch. Gesam. Virusforsch.* **33,** 288.
Krishna, R. V., Meurman, O. H., Ziegler, T., and Krech, U. H. (1980). *J. Clin. Microbiol.* **12,** 46.
Ksiazek, T. G., Olson, J. G., Irving, G. S., Settle, C. S., White, R., and Petrusso, R. (1980). *Am. J. Epidemiol.* **112**(4), 487.
Kuhlmann, W. D. (1979). *Histochemistry* **64,** 67.
Kuhlmann, W. D. (1984). "Immuno Enzyme Techniques in Cytochemistry." Verlag Chemie International. Deerfield Beach, Florida.
Kung, P. C., Lang, J. C., McCaffre, R. P., Ratcliff, R. L., Harrison, T. A., and Baltimore, D. (1978). *Am. J. Med.* **64,** 788.
Kuno, G., Gubler, D. J., and Santiago De Weil, S. (1985). *J. Virol. Methods* **12,** 93.
Kuno-Sakai, H., Iwasaki, H., and Kumura, M. (1984). *J. Infect. Dis.* **149,** 813.
Kurstak, C., Morisset, R., and Kurstak, E. (1982). *Ann. Virol. (Inst. Pasteur)* **133E,** 187.
Kurstak, E. (1971). *In* "Methods in Virology" (K. Maramorosch and H. Koprowski, eds.), Vol. V, pp. 423–444. Academic Press, New York.
Kurstak, E. (1985). *Bull. WHO* **63**(4), 793–811.
Kurstak, E., and Kurstak, C. (1974). *In* "Viral Immunodiagnosis" (E. Kurstak and R. Morisset, eds.), pp. 1–30. Academic Press, New York.
Kurstak, E., Côté, J. R., and Belloncik, S. (1969). *C.R. Acad. Sci. Paris* **268,** 2309.
Kurstak, E., Belloncik, S., and Garzon, S. (1971). *Curr. Top. Microbiol. Immunol.* **55,** 200.
Kurstak, E., Belloncik, S., Onjii, P. A., Montplaisir, S., and Martineau, B. (1972). *Arch. Ges. Virus* **38**(1), 67.
Kurstak, E., Viens, P., Morisset, R., Kurstak, C., and Charpentier, G. (1973). *C.R. Acad. Sci. Paris* **276,** 2345.

Kurstak, E., Tijssen, P., Kurstak, C., and Morisset, R. (1975). *Ann. New York Acad. Sci.* **254,** 369.

Kurstak, E., Tijssen, P., and Kurstak, C. (1977). *In* "Comparative Diagnosis of Viral Diseases" (E. Kurstak and C. Kurstak, eds.), Vol. II, pp. 403–488. Academic Press, New York.

Kurstak, E., de Thé, G., van den Hurk, J., Charpentier, G., Kurstak, C., Tijssen, P., and Morisset, R. (1978). *J. Med. Virol.* **2,** 189.

Kurstak, E., Tijssen, P., van den Hurk, J., Kurstak, C., and Morisset, R. (1980). *In* "Invertebrate Systems in Vitro" (E. Kurstak, K. Maramorosch, and A. Dübendorfer, eds.), pp. 365–373. Elsevier Biomedical Press, Amsterdam.

Kurstak, E., Kurstak, C., van den Hurk, J., and Morisset, R. (1981). *In* "Comparative Diagnosis of Viral Diseases" (E. Kurstak and C. Kurstak, eds.), Vol. IV, pp. 105–148. Academic Press, New York.

Kurstak, E., Kurstak, C., and Morisset, R. (1982). *Stand. Immunol. Proc. Dev. Biol. Stand.* **52,** 187.

Kurstak, E., Tijssen, P., and Kurstak, C. (1984a). *In* "Control of Virus Diseases" (E. Kurstak and R. Marusyk, eds.), pp. 477–500. Dekker, New York.

Kurstak, E., Tijssen, P., and Kurstak, C. (1984b). *In* "Applied Virology" (E. Kurstak, ed.), pp. 479–503. Academic Press, New York.

Kurstak, E., Tijssen, P., Kurstak, C., and Morisset, R. (1986). *Bull. WHO* **64**(3)(A. 4450).

Kurtz, D. T., and Nicodemus, C. F. (1981). *Gene* **13,** 145.

Kurtz, J. B., and Malic, A. (1981). *J. Clin. Pathol.* **34,** 1392.

Kusama, H. (1983). *J. Clin. Microbiol.* **17,** 317.

Lachmann, P. J., Oldroyd, R. G., Wright, B., and Milstein, C. (1980). *J. Immunol.* **124,** 1527.

Laflamme, M. Y., Kurstak, C., Kurstak, E., and Morisset, R. (1976). *Can. J. Ophthal.* **11,** 217.

Lamoyi, E., and Nisonoff, A. (1983). *J. Immunol. Methods* **56,** 235.

Lander, J. J., Alter, H. J., and Purcell, R. H. (1971). *J. Immunol.* **106,** 1166.

Langone, J. J. (1982). *Adv. Immunol.* **32,** 157.

Langone, J. J., and Van Vunakis, H., eds. (1981). *Methods Enzymol.* **74,** 1–729.

Langone, J. J., and Van Vunakis, H., eds. (1983). *Methods Enzymol.* **92,** 1–647.

Lanzillo, J. J., Stevens, J., Tumas, J., and Fanbury, B. L. (1983). *Electrophoresis* **4,** 313.

Laver, W. G. (1984). "Molecular Immunology," p. 142. Dekker, New York.

Laver, W. G., Air, G. M., Webster, R. G., Gerhard, W., Ward, C. W., and Depheide, T. A. A. (1979). *Virology* **98,** 226.

Leary, J. J., Brigati, D. J., and Ward, D. C. (1983). *Proc. Natl. Acad. Sci. U.S.A.* **80,** 4045.

Leder, P. (1982). *Sci. Am.* **246** 102.

Leggiadro, R., Yolken, R. H., Simkins, J. H., and Hughes, W. T. (1982). *J. Infect. Dis.* **144,** 484.

Lehtonen, O.-P., and Eerola, E. (1982). *J. Immunol. Methods* **54,** 233.

Leinikki, P. O., and Passila, S. (1977). *J. Infect. Dis.* **136** (Suppl.) S 294.

Leinikki, P. O., Shekarcki, I., Dorsett, P., and Sever, J. L. (1978). *J. Lab. Clin. Med.* **92,** 849.

Leinikki, P. O., Shekarcki, I., Tzan, N., Madden, D. L., and Sever, J. L. (1979). *Proc. Soc. Exp. Biol. Med.* **160,** 363.

Levy, H. B., and Sober, H. A. (1960). *Proc. Soc. Exp. Med. Biol.* **103,** 250.

Lewis, G. E., Kulinski, S. S., Reichard, D. W., and Metzger, J. F. (1981). *Appl. Environ. Microbiol.* **42,** 1018.

Lin, W., and Kasamatsu, H. (1983). *Anal. Biochem.* **128,** 302.

Lindgren, J., Wahlström, T., Bang, B., Hurme, M., and Mäkelä, O. (1982). *Histochemistry* **74,** 223.

Livingstone, D. M. (1974). *Methods Enzymol.* **34,** 723.

Locarnini, S. A., Coulepis, A. G., Stratton, A. M., Kaldor, J., and Gust, I. D. (1979). *J. Clin. Microbiol.* **9,** 459.

Lyerly, D. M., Sullivan, N. M., and Wilkins, T. D. (1983). *J. Clin. Microbiol.* **17,** 72.

McClane, B. A., and Strouse, R. J. (1984). *J. Clin. Microbiol.* **19,** 112.

McCullough, K. C., Crowther, J. R., and Butcher, R. N. (1985). *J. Virol. Methods* **11,** 329.

McCullough, N. B. (1976). "Immune-response to Brucella M: Manual of Clinical Immunology" (N. R. Rose and H. Friedman, eds.), American Society of Microbiology, Washington, D.C.

McIntosh, K. (1983). *Curr. Top. Microbiol. Immunol.* **104,** 1.

McIntosh, K., Hendry, R. M., Fahnestock, M. L., and Pierik, L. T. (1982). *J. Clin. Microbiol.* **16,** 329.

Macky, L. J., McGregor, F. A., Paaunova, N., and Lambert, P. H. (1982). *Bull. WHO* **60,** 69.

McLaren, M. L., Lillywhite, J. E., Dunne, D. W., and Doenhoff, M. J. (1981). *Trans. R. Soc. Trop. Med. Hyg.* **75,** 72.

McLean, B., Sonza, S., and Holmes, I. H. (1980). *J. Clin. Microbiol.* **12,** 314.

McLean, D. M. (1979). *In* "Arctic and Tropical Arboviruses" (E. Kurstak, ed.), pp. 7–19. Academic Press, New York.

McLean, D. M., Grass, P. N., Judd, B. D., Stolts, R. J., and Wong, K. K. (1978). *Arch. Virol.* **55,** 315–322.

MacLean, I. W., and Nakane P. K. (1974). *J. Histochem. Cytochem.* **22,** 1077.

McMichael, J. C., Greisiger, L. M., and Millman, I. (1981). *J. Immunol. Methods* **45,** 79.

McMillan, E. M., Wasik, R., and Everett, M. A. (1981). *Am. J. Clin. Pathol.* **76,** 737.

McMillan, E. M., Wasik, R., and Everett, M. A. (1982). *Arch. Pathol. Lab. Med.* **106,** 9.

Macnaughton, M. R., Flowers, D., and Isaacs, D. (1983). *J. Med. Virol.* **11,** 319.

Maddison, S. E., Hayes, G. V., Slemenda, S. B., Norman, L. G., and Ivey, M. H. (1982). *J. Clin. Microbiol.* **15,** 1036.

Madore, H. P., Reichman, R. C., and Dolin, R. (1983). *J. Clin. Microbiol.* **18,** 1345.

Maeda, M., Ito, K., Arakawa, H., and Tsuji, A. (1985). *J. Immunol. Methods* **82,** 83.

Mahony, J. B., Schachter, J., and Chernesky, M. A. (1983). *J. Clin. Microbiol.* **18,** 270.

Major, P. P., Morisset, R., Kurstak, C., and Kurstak, E. (1978). *C.M.A. J.* **8**(118), 821.

Malvano, R., Moniolo, A., Davis, M., and Zannino, M. (1982). *J. Immunol. Methods* **48,** 51.

Mardiney, M. R., Muller-Eberhad, N., and Feldman, J. D. (1968). *Am. J. Pathol.* **53,** 253.

Marucci, A. A., and Dougherty, R. M. (1975). *J. Histochem. Cytochem.* **23,** 618.

Marusyk, R. G., and Tyrrell, D. L. J. (1984). *In* "Control of Virus Disease" (E. Kurstak and R. G. Marusyk, eds.), pp. 67–78. Dekker, New York.

Mason, D. Y., and Sammons, R. (1978). *J. Clin. Pathol.* **31,** 454.

Mason, D. Y., and Sammons, R. (1979). *J. Histochem. Cytochem.* **27,** 832.

Mason, D. Y., and Woolston, R. E. (1982). *In* "Techniques in Immunochemistry" (G. R. Bullock and P. Petrusz, eds.), Vol. 1, p. 125. Academic Press, London.

Mason, D. Y., Aldulaziz, Z., Falini, B., and Stein, H. (1983). *In* "Immunochemistry: Practical Applications in Pathology and Biology" (J. M. Polak and S. Van Noorden, eds.), p. 113. Wright PSG, Bristol.

Mason, T. E., Phifer, R. F., Spios, S. S. Swallow, R. A., and Dreskin, R. B. (1969). *J. Histochem. Cytochem.* **17,** 563.

Matthews, J. B. (1981). *J. Clin. Pathol.* **34,** 103.

Matthews, H. M., Walls, K. W., and Huang, A. Y. (1984). *J. Clin. Microbiol.* **19,** 221.

Maze, M., and Gray, G. M. (1980). *Biochemistry* **19,** 2351.

Meisels, A. and Morin, C. (1981). *Gynecol. Oncol.* **12,** S 111.

Mellors, R. C., Ortega, L. G., and Holmen, H. R. (1957). *J. Exp. Med.* **106,** 191.

Mendelsohn, G., Eggleston, J. C., and Mann, R. B. (1980). *Cancer* **45,** 273.

Mersels, A., and Morin, C. (1981). *Gynecol. Oncol.* **12,** 5111.

Meurman, O. H., and Ziola, B. R. (1978). *J. Clin. Pathol.* **31,** 483.

Meurman, O. (1983). *Curr. Top. Microbiol. Immunol.* **104,** 101.

Meurman, O., Sarkkinen, H., Ruuskanen, O., Hänninen, P., and Halonen, P. (1984a). *J. Med. Virol.* **14,** 61.

Meurman, O., Ruuskanen, O., Sarkkinen, H., Hänninen, P., and Halonen, P. (1984b). *J. Med. Virol.* **14,** 67.

Mills, T., Cunningham, S., Alvers, D., and Harrison, J. (1982). *Dev. Biol. Stand.* **52,** 367.

Milstein, C. (1982). *Cancer* **49,** 1953.

Mirsky, R., Winter, J., Abney, E. R., Pruss, R. M., and Gavrilov, Y. (1980). *J. Cell Biol.* **84,** 483.

Mitchell, G. F., and Anders, R. F. (1982). *Antigens* **6,** 69.

Mitchell, G. F., Premier, R. R., Garcia, E. G., Hurrell, J. GR. Chandler, H. M., Cruise, K. M., Tapales, F. E. P., and Tiu, W. U. (1983). *Am. J. Trop. Med. Hyg.* **32,** 114.

Modesto, R. R., and Pesce, A. J. (1971). *Biochim. Biophys. Acta* **229,** 384.

Molin, S.-O., Nygren, H., and Dononius, L. (1978). *J. Histochem. Cytochem.* **26,** 1053.

Morgan-Capner, P., Tedderm, R. S., and Mace, J. E. (1983). *J. Hyg.* **90,** 407.

Moriarty, G. C., and Holmi, N. S. (1972). *J. Histochem. Cytochem.* **20,** 590.

Morisset, R., Kurstak, C., and Kurstak, E. (1974). *In* "Viral Immunodiagnosis" (E. Kurstak and R. Morisset, eds.), pp. 31–39. Academic Press, New York.

Morisset, R., and Kurstak, E. (1985). "Advances in Sexually Transmitted Diseases. Treatment and Diagnosis." VNU Science Press, Utrecht, Holland.

Morrison, D. C., and Leive, L. (1975). *J. Biol. Chem.* **250,** 2911.

Morrow, D. L., Kline, J. B., Douglas, S. D., and Polin, R. A. (1984). *J. Clin. Microbiol.* **19,** 400.

Moskophidis, M., and Müller, F. (1985). *Eur. J. Clin. Microbiol.* **4**(5), 473.

Mössner, E., Boll, M., and Pfleiderer, G. (1980). *Hoppe Seyler's Z. Physiol. Chem.* **361,** 543.

Mott, K. E., and Dixon, H. (1982). *Bull. WHO* **60,** 729.

Moynihan, M., and Petersen, I. (1981). *Dev. Biol. Stand.* **47,** 129.

Mufson, M. A. (1978). *In* "Handbook in Clinical Laboratory Sciences" (D. Seligson, ed.), Vol. 1, p. 201. CRC Press, West Palm Beach, Florida.

Mulé, S. J., Jikofsky, D., Kogan, M., DePace, A., and Vereberg, K. (1977). *Clin. Chem.* **23,** 796.

Mumford, R. S., and Hall C. L. (1979). *Infect. Immun.* **26,** 42.

Murakami, S. S., and Said, J. W. (1984). *Am. J. Clin. Pathol.* **81,** 293.

Muraro, R., Wunderlich, D., Thor, A., Lundy, J., Noguchi, P., Cunningham, R., and Schlom, J. (1985). *Cancer Res.* **45,** 5769.

Mushahwar, I. K., Gerin, J. L., Dienstag, J. L., Decker, R. G., Smedile, A., and Rizzetto, M. (1984). *Pathologists* **38,** 10.

Murphy, B. R., Phelan, M. A., Nelson, D. L., Yarchoa, R., Tierney, E., Alleng, D. W., and Chanock, R. M. (1981). *J. Clin. Microbiol.* **13,** 554.

Nadji, M. (1981). *Urol. Times* **9,** 6.

Nadji, M., Tabei, S., Casho, A., Chu, T., and Morales, A. R. (1980). *Am. J. Clin. Pathol.* **73,** 735.

Nahmias, A. J., Delcuoni, I., Pipkin, J., Hutton, R., and Wickliffe (1971). *Appl. Microbiol.* **22,** 455.

Nairn, R. C. (1976). "Fluorescent Protein Tracing," 4th ed. Churchill Livingstone, London.

Nakane, P. K., and Farr, A. G. (1981). *J. Immunol. Methods* **47,** 129.

Nakane, P. K., and Kawaoi, A. (1974). *J. Histochem. Cytochem.* **22,** 1084.

Nakane, P. K., and Pierce, G. B. (1966). *J. Histochem. Cytochem.* **14,** 929.

Nakata, S., Chiba, S., Terashima, M., Sakuma, Y., Kogasaka, R., and Nakao, T. (1983). *J. Clin. Microbiol.* **17,** 198.

Naot, Y., and Remington, J. S. (1980). *J. Infect. Dis.* **142,** 757.

Nash, T. E., Gracia-Cozco, C., Ruiz-Tiben, E., Nazario-Lopez, H. A., Vasquez, G., and Torres-Borges, A. (1983). *Am. J. Trop. Med.* **32,** 776.

Nathwani, B. N. (1979). *Cancer* **44,** 347.

Neurath, A. R., and Strick, N. (1981). *J. Virol. Methods* **3,** 155.

Neurath, A. R., Strick, N., Lee, Y. S., Nilsen, T., Baker, L., Sproul, P., Rubenstein, P., Taylor, P., Stevens, C. E., and Gold, J. W. M. (1985). *J. Virol. Methods* **12,** 85.

Newhall, W. J., Batteiger, B., and Jones, R. B. (1982). *Infect. Immun.* **38,** 1181.

Ngo, T. T., and Lenhoff, H. M. (1981). *Biochem. Biophys. Res. Commun.* **99,** 496.

Nicolai-Scholten, M. E., Ziegelmares, R., Behrens, F., and Hopken, W. (1980). *Med. Microbiol. Immunol.* **168,** 81.

Nicolas, J. C., Cohen, B., Fortier, B., Lourenco, M. H., and Bricout, P. (1983). *Virology* **124,** 181.

Nielsen, P. J., Manchester, K. L., Towbin, H., Gordon, J., and Thomas, G. (1982). *J. Biol. Chem.* **257,** 12316.

Nisonoff, A., Hopper, J. E., and Spring, S. B. (1975). "The Antibody Molecule." Academic Press, New York.

Nomura, M., Imai, M., Usuda, S., Nakamura, T., Miyakawa, Y., and Mayumi, M. (1983). *J. Immunol. Methods* **56,** 13.

Norrby, E. (1978). *Prog. Med. Virol.* **24,** 1.

Nygren, H. (1982). *J. Histochem. Cytochem.* **30,** 407.

O'Brien, M. J., Zamcheck, N., Burke, B., Kirkham, S. E., Saravis, C. A., and Gottlieb, L. S. (1981). *Am. J. Clin. Pathol.* **75,** 283.

O'Connor, C. G., and Ashman, L. K. (1982). *J. Immunol. Methods* **54,** 267.

Oellerich, M., Sybrecht, G. W., and Haeckel, R. (1979). *J. Clin. Chem. Clin. Biochem.* **17,** 299.

Oellerich, M., Haeckel, R., and Haindl, H. (1982). *J. Clin. Chem. Clin. Biochem.* **20,** 765.

Oliver, D. G., Sanders, A. H., Hogg, R. D., and Hellman, J. W. (1981). *J. Immunol. Methods* **42,** 195.

Oliver-Gonzalez, J. (1954). *J. Infect. Dis.* **95,** 86.

Olmsted, J. B. (1981). *J. Biol. Chem.* **256,** 11955.

Ordronneau, P. (1982). *In* "Techniques in Immunocytochemistry" (G. R. Bullock and P. Petrusz, eds.), Vol. 1, p. 269. Academic Press, London.

Osterhaus, A. D. M. R., van Wezel, A. L., Hazendont, T. G., WytedeHaag, F. G. C. M., van Asten, J. A. A. M., and van Stenis, B. (1983). *Intervirology* **20,** 129.

O'Sullivan, M. J., Gnemmi, E., Morris, D., Chieregatto, G., Simmons, M., Simmonds, M., Bridges, J. W., and Marks, V. (1979). *J. Immunol. Methods* **30,** 127.

Painter, K., and Vader, C. R. (1979). *Clin. Chem.* **25,** 797.

Palfree, R. G. E., and Elliott, B. E. (1982). *J. Immunol. Methods* **52,** 395.

Papasian, C. J., Bartholomeo, W. R., and Amsterdam, D. (1984). *J. Clin. Microbiol.* **19,** 347.

Parkinson, A. J., Scott, E. N., and Muchmore, H. G. (1982). *J. Clin. Microbiol.* **15,** 538.

Partonen, P., Turunen, H. J., Paasivuo, R., Forsblem, L., Suni, J., and Leinikki, P. O. (1983). *FEBS Lett.* **158,** 252.

Pauling, L., Pressman, D., and Grossberg, A. L. (1944). *J. Am. Chem. Soc.* **66,** 784.

Paxton, J. W. (1979). *Clin. Chem.* **25,** 491.

Payne, P. A., Isaac, M., and Francis, J. M. (1982). *J. Clin. Pathol.* **35,** 892.

Pearse, A. G. E. (1980). "Histochemistry, Theoretical and Applied," 4th ed. Churchill Livingstone, Edinburgh.

Peferoen, M., Huybrechts, R., and DeLoof, A. (1982). *FEBS Lett.* **145,** 369.

Peltre, G. (1982). *J. Allergy Clin. Immunol.* **69,** 111.

Perera, V. Y., Creasy, M., and Winter, A. J. (1983). *J. Clin. Microbiol.* **18,** 601.

Peterson, A. (1981). *J. Biol. Chem.* **256,** 6975.

Peterson, E. A. (1970). *In* "Laboratory Techniques in Biochemistry and Molecular Biology" (T. S. Work and E. Work, eds.), Vol. 2, pt. 2. Elsevier North Holland, Amsterdam.

Petit, C., Sauron, M. E., Gilbert, M., and Thexe, J. (1982). *Ann. Immunol.* **133,** 77.

Phaneuf, D., Francke, E. L., and Neu, H. C. (1980). *Curr. Chemother. Inf. Dis.* **1,** 512.

Phelan, M. A., Mayner, R. E., Bucher, D. J., and Ennis, F. A. (1980). *J. Biol. Stand.* **8,** 233.

Pickel, V. M., Joh, T. H., and Reis, D. J. (1976). *J. Histochem. Cytochem.* **24,** 792.

Picq, J. J., Faugere, B., Sebeco, D., and Peney, J. M. (1979). *Bull. Soc. Pathol. Exat.* **72,** 231.

Pifer, L. L., Hughes, W. T., Stagno, S., and Woods, D. (1978). *Pediatrics* **61,** 35.

Pillai, S., and Mohimen, A. (1982). *Gastroenterology* **83,** 1210.

Pinon, J. M., Sulakion, A., Remy, G., and Dropsy, G. (1979). *Am. J. Trop. Med. Hyg.* **28,** 318.

Pinon, J. M., Thoannes, H., and Gruson, N. (1985). *J. Immunol. Methods* **77,** 15.

Plattner, H., Wachter, E., and Gröbner, P. (1977). *Histochemistry* **53,** 223.

Polak, J. M., and van Noorden, S., eds. (1983). Immunochemistry: Practical Applications in Pathology and Biology." Wright PSG, Bristol.

Polin, R. A., and Kennett, R. (1980). *J. Clin. Microbiol.* **11,** 332.

Ponder, B. A., and Wilkinson, M. M. (1981). *J. Histochem. Cytochem.* **29,** 981.

Popano-Krauff, T. (1981). *J. Med. Virol.* **8,** 79.

Porter, D. D., and Porter, H. G. (1984). *J. Immunol. Methods* **72,** 1.

Portsmann, B., Avrameas, S., Ternynck, T., Portsmann, T., Michell, B., and Guesdon, J.-L. (1984). *J. Immunol. Methods* **66,** 179.

Pouletty, Ph., Kadouche, J., Garcia-Gonzalez, M., Mitraesco, E., Desmonts, G., Thulliez, Ph., Thoannes, H., and Pinon, J. M. (1985). *J. Immunol. Methods* **76,** 289.

Prat, M., Morra, I., Bussolati, G., and Comoglio, P. (1985). *Cancer Res.* **45,** 5799.

Primus, F. J., Clark, C. A., and Goldenberg, D. M. (1982). *J. Natl. Cancer Inst.* **67,** 1031.

Pugh, S. F., Slack, R. C. B., Caul, E. O., Paul, I. D., Appleton, P. N., and Gatley, S. (1985). *J. Clin. Pathol.* **38,** 1139.

Purcell, R. H., Guerin, J. L., Almeida, J. D., and Holland, P. V. (1973). *Intervirology* **2,** 231.

Radaszkiewicz, T., Drgosics, B., Abdelfattahgad, M., and Denk, H. (1979). *J. Immunol. Methods* **29,** 29.

Rakela, J., and Mosley, J. W. (1977). *J. Infect. Dis.* **139,** 933.

Randle, B. J., and Epstein, M. A. (1984). *J. Virol. Methods* **9,** 201.

Ranki, A., Reitamo, S., Kanttinen, T., and Hayry, P. (1980). *J. Histochem. Cytochem.* **28,** 704.

Rapoport, H. (1966). "Atlas of Tumor Pathology." Armed Forces Institute of Pathology, Washington, D.C.

Rawls, W. E., Clarke, A., Smith, O., Docherty, J. J., Gilman, S. C., and Graham, S. (1980). *Cold Spring Harbor Conf. Cell Prolif.* **7,** 117.

Raymond, J., Duc-Goiron, P., Joundy, S., Orfila, J., and Acar, J. (1985). *Eur. J. Clin. Microbiol.* **4**(5), 468.

Reedman, B. M., and Klein, G. (1973). *Int. J. Cancer* **11,** 499.

Reeve, P., Owen, J., and Oreil, J. D. (1975). *J. Clin. Pathol.* **28,** 910.

Reeves, W. G., Allen, B. R., and Tatersall, R. B. (1980). *Br. Med. J.* **280,** 1500.

Reeves, W. G., Corey, L., Adams, H. G., Vantner, A., and Holmes, K. K. (1981). *New Engl. J. Med.* **305,** 315.

Reimer, C. B., Black, C. M., Phillips, D. J., Logan, L. C., Hunter, E. F., Pender, B. J., and McGrew, B. E. (1977). *Ann. N.Y. Acad. Sci.* **254,** 77.

Reimer, C. B., Phillips, D. J., Black, C. M., and Wells, T. W. (1978). *In* "Immunofluorescence and Related Staining Techniques" (W. Knapp, H. Holubar, and G. Wick, eds.), p. 189. Elsevier/North Holland, Amsterdam.

Reiner, M., and Wecker, E. (1981). *Med. Microbiol. Immunol.* **169,** 237.

Reinhart, M. P., and Malamud, D. (1982). *Anal. Biochem.* **123,** 229.

Reiser, J., and Wardale, J. (1981). *Eur. J. Biochem.* **114,** 569.

Reiser, R., Conaway, D., and Bergdoll, M. S. (1974). *Appl. Microbiol.* **27,** 83.

Renart, J., Reiser, J., and Stark, G. R. (1979). *Proc. Natl. Acad. Sci. U.S.A.* **76,** 3116.

Reno, D. S., Andersen, S. G., and Grab, B. (1970). *Bull. WHO* **42,** 535.

Richard-Lenoble, D., Smith, M. D., and Loisy, R. (1978). *Ann. Trop. Med. Parasitol.* **72,** 553.

Richardson, M. D., Turner, A., Warnock, D. W., and Llewellyn, P. A. (1983). *J. Immunol. Methods* **56,** 201.

Rigby, P. W. J., Dreckmann, M., Rhodes, C., and Berg, P. (1977). *J. Mol. Biol.* **113,** 237.

Rimondo, G., Lonto, G., and Squadrito. (1983). *Br. Med. J.* **286,** 845.

Rittenhouse, H. G., Petruska, J. D., and Hirata, A. A. (1984). *Prot. Biol. Fluids* **31,** 937.

Rizzetto, M. (1983). *Hepatology* **3,** 729.

Rizzetto, M., Canese, M. G., Arico, S., Crivelli, O., Trepo, C., Bonino, F., and Verme, G. (1977). *Gut* **18,** 997.

Rodbard, D., and McClean, S. W. (1977). *Clin. Chem.* **23,** 112.

Rodger, S. M., Bishop, R. F., and Holmes, I. H. (1982). *J. Clin. Microbiol.* **16,** 724.

Rodgers, R. P. C. (1984). *In* "Practical Immunoassay. The State of Art" (W. R. Butt, ed.), p. 253. Dekker, New York.

Rodning, C. B., Erlandsen, S. L., Coulter, H. D., and Wilson, I. D. (1980). *J. Histochem. Cytochem.* **26,** 223.

Rodriguez, R. L., and Tait, R. C. (1983). "Recombinant DNA Techniques." Addison-Wesley, London.

Roseto, A., Vautherat, J. F., Bobulesco, P., and Guillemin, M. C. (1982). *C.R. Acad. Sci. Paris* **294,** 347.

Rowe, D. S., and Fahey, J. L. (1965). *J. Exp. Med.* **121,** 171.
Rowe, D. S., Anderson, S. G., and Grub, B. (1970). *Bull. WHO* **43,** 607.
Rowley, G. L., Rubenstein, K. E., Huisken, J., and Ullman, E. F. (1975). *J. Biol. Chem.* **250,** 3759.
Rubenstein, K. E., Schneider, R. S., and Ullman, E. F. (1972). *Biochem. Biophys. Res. Commun.* **47,** 846.
Rubenstein, A. S., Chan, R., Ling, C., Miller, M., Nehmadi, F., and Overby, L. (1980). *Ann. Meeting Am. Soc. Microbiol.* (Abstract).
Ruitenberg, E. J., and Brosi, B. J. M. (1978). *Scand. J. Immunol.* **8**(7), 63.
Ruitenberg, E. J., and Buys, J. (1977). *Am. J. Trop. Med. Hyg.* **26,** 31.
Ruitenberg, E. J., and van Knapen, F. (1977). *J. Infect. Dis.* **136,** S261.
Ruitenberg, E. J., Brosi, B. J. M., and Steerenberg, P. A. (1976). *J. Clin. Microbiol.* **3,** 541.
Ruiz-Tiben, E., Hillyer, G. V., Knight, W. B., Gomez des Rios, I., and Woodall, J. P. (1979). *Am. J. Trop. Med. Hyg.* **28,** 230.
Ryan, J. W., Day, A. R., Schutz, D. R., Ryan, U. S., Chung, A., Marborough, D. I., and Dorer, F. E. (1976). *Tissue Cell* **8,** 111.
Rybicki, E. P., and von Wechmar, M. B. (1982). *J. Virol. Methods* **5,** 267.
Said, J. W., Nash, G., and Lee, M. (1982). *Human Pathol.* **13,** 1106.
Saikku, P., Paavonen, J., Väänänen, P., and Vaheri, A. (1983). *J. Clin. Microbiol.* **17,** 22.
Sainte-Marie, G. (1962). *J. Histochem. Biochem.* **10,** 250.
Sakata, H., Hishiyama, M., and Sigiura, A. (1984). *J. Clin. Microbiol.* **19,** 21.
Sakuma, Y., Chiba, S., Kogasaka, R., Terashima, H., Nakamura, S., Horizo, K., and Nakoo, T. (1981). *J. Med. Virol.* **7,** 221.
Salk, J., Cohen, H., Fillastre, C., Stoeckel, P., Rey, J. L., Schlumberger, M., Nicolas, A., van Stenis, G., van Wezel, A. L., Triau, R., Saliou, P., Barry, L. F., Moreau, J.-P., and Mérieux, C. (1978). *Dev. Biol. Stand.* **41,** 119.
Sarkkinen, H. K., Halonen, P. E., Arstila, P. P., and Salmi, A. A. (1981a). *J. Clin. Microbiol.* **13,** 258.
Sarkkinen, H. K., Halonen, P. E., and Salmi, A. A. (1981b). *J. Med. Virol.* **7,** 213.
Sarkku, F., Paavonen, J., Väänänen, P., and Vaheri, A. (1983). *J. Clin. Microbiol.* **17,** 22.
Sarov, I., and Haikin, H. (1983). *J. Virol. Methods* **6,** 161.
Saunders, G. C., Clinard, E. H., Bartlett, M. L., and Mort Sanders, W. M. J. (1977). *J. Infect. Dis.* **130,** S258.
Schachter, J., Cles, Z., Ray, R., and Hines, P. A. (1979). *J. Clin. Microbiol.* **10,** 647.
Schachter, J., Grossman, M., and Azimi, P. (1982). *J. Infect. Dis.* **146,** 530.
Schachter, J., McCormack, W. M., Smith, R. F., Parks, R. M., Bailey, R., and Ohlin, K. K. (1984). *J. Clin. Microbiol.* **19,** 57.
Schantz, P. M., Shanks, D., and Wilson, M. (1980). *Am. J. Trop. Med. Hyg.* **29,** 609.
Scheifele, D. W., Ward, J. I., and Silber (1981). *Pediatrics* **68,** 888.
Schild, G. C. (1984). *In* "Control of Virus Diseases" (E. Kurstak and R. G. Marusyk, eds.), pp. 93–114. Dekker, New York.
Schiller, E. L. (1967). *Adv. Clin. Chem.* **9,** 43.
Schlegel, R., Banks-Schlegel, S., and McLeod, J. A. (1980). *Am. J. Pathol.* **101,** 41.
Schluederberg, A. (1965). *Nature (London)* **205,** 1232.
Schmidt, N. J., Ho, H. H., and Chin, J. (1981). *J. Clin. Microbiol.* **13**(4), 627.
Schmidt, O. W., and Kenny, G. E. (1982). *Infect. Immun.* **35,** 515.
Schmidt, O. W. (1984). *J. Clin. Microbiol.* **20,** 175.
Schmitz, H., Doerr, H. W., Kampa, D., and Vogt, A. (1977). *J. Clin. Microbiol.* **5,** 629.

Schmitz, H., von Deimling, U., and Flehmig, B. (1980a). *J. Gen. Virol.* **50,** 59.
Schmitz, H., von Deimling, U., and Flehmig, B. (1980b). *J. Infect. Dis.* **142,** 250.
Schofield, L., Saul, A., Myler, P., and Kidsen, C. (1982). *Infect. Immun.* **38,** 893.
Schottelius, D. D. (1978). *In* "Antiepileptic Drugs" (C. E. Pippenger, J. K. Perry, and H. Kutt, eds.). Raven, New York.
Schroeder, L. L. (1985). *J. Immunol. Methods* **83,** 135.
Schuurs, A. H. W. M., and Wolters, G. (1975). *Am. J. Med. Sci.* **270,** 173.
Scoggin, D., Petrekn, J., and Besemer, D. (1978). *Clin. Chem.* **24,** 1055.
Sedgwick, A. K., Ballow, M., Sparks, K., and Tilton, R. C. (1983). *J. Clin. Microbiol.* **18,** 104.
Sever, J. L. (1983). *Curr. Top. Microbiol. Immunol.* **104,** 57.
Sharkey, R. M., Primus, F. J., and Goldenberg, D. M. (1980). *Histochemistry* **66,** 35.
Sharma, S. D., Mullenan, J., Araujo, F. G., Erlich, H. A., and Remington, J. S. (1983). *J. Immunol.* **131,** 977.
Shattock, A. G., and Fielding, J. F. (1983). *Br. Med. J.* **286,** 1279.
Shattock, A. G., and Morgan, B. M. (1984). *J. Med. Virol.* **13,** 73.
Shekarchi, I. C., Sever, J. L., Ward, L. A., and Madden, D. L. (1982). *J. Clin. Microbiol.* **16,** 1012.
Sheridan, J. F., Aurelian, L., Barbour, G., Santosham, M., Sack, R. B., and Ryder, R. W. (1981). *Infect. Immun.* **31,** 419.
Sinclair, R. A., Burns, J., and Dunnill, M. S. (1981). *J. Clin. Pathol.* **34**(8), 859.
Singer, Y., Kimmel, N., and Sarov, I. (1985). *J. Virol. Methods* **11,** 29.
Sippel, J. E., and Voller, A. (1980). *Trans. R. Soc. Trop. Med. Hyg.* **74,** 644.
Siskind, G. W. (1973). *Pharm. Rev.* **25,** 319.
Skelly, J., Howard, C. R., and Zuckerman, A. J. (1978). *J. Gen. Virol.* **41,** 447.
Skinhøj, P., Mikkelsen, F., and Hollinger, F. B. (1977). *Am. J. Epidemiol.* **105,** 140.
Skurrie, I. J., and Gilbert, G. L. (1983). *J. Clin. Microbiol.* **17,** 738.
Slighton, E. L. (1978). *J. For. Sci.* **23,** 292.
Sloane, J. P., and Ormerod, M. G. (1981). *Cancer* **47,** 1786.
Smedile, A., Lavarini, C., Farci, P., Arico, S., Marinucci, G., Dentico, P., Cargnel, A., Opolen, P., Blanco, C., and Rizetto, M. (1983). *Am. J. Epidemiol.* **117,** 223.
Smeltzer, M. P., Lossick, J. G., and Elliott, K. M. (1985). *In* "Advances in Sexually Transmitted Diseases" (R. Morisset and E. Kurstak, eds.), pp. 75–85. VNU Science Press, Utrecht, Holland.
Smith, P. D., Gillin, F. D., Brown, W. R., and Nash, T. E. (1981). *Gastroenterology* **80,** 1476.
Smith, I. W., Peutherer, J. F., and Robertson, D. H. H. (1976). *Lancet* **2,** 1089.
Soula, A., Moreau, Y., Laurent, N., and Tixier, G. (1981). *Dev. Biol. Stand.* **52,** 147.
Soulebot, J. P., Brun, A., and Dubourget, Ph. (1982). *Dev. Biol. Stand.* **52,** 451.
Speers, W. G., Picaso, L. G., and Silverberg, S. G. (1983). *Am. J. Clin. Pathol.* **79,** 105.
Spencer, E., Avendano, F., and Aroya, M. (1983). *J. Infect. Dis.* **148,** 41.
Spencer, H. C., Allain, D. S., Sulzen, A. J., and Collins, W. E. (1980). *Am. J. Trop. Med. Hyg.* **29,** 179.
Stallman, N. D., Allan, B. C., and Sutherland, C. J. (1974). *Med. J. Aust.* **3,** 629.
Starling, J. J., Sieg, S. M., Beckett, M. L., Wirth, P. R., Wahab, Z., Schellhammer, P. F., Ladaga, L. E., Poleskic, S., and Wright, G. L. (1986). *Cancer Res.* **46,** 367.
Stamm, W. E., Cole, B., Fennell, C., Bonin, P., Armstrong, J. E., Hermann, J. E., and Holmes, K. K. (1984). *J. Clin. Microbiol.* **19,** 57.

Stass, S. A., Dean, L., Perper, S. C., and Bollum, F. J. (1982). *Am. J. Clin. Pathol.* **77,** 174.

Stefanini, M., De Martino, C., and Zamboni, L. (1967). *Nature (London)* **216,** 173.

Stein, B. S., Peterson, R. O., Vangere, S., and Kendall, A. R. (1982). *Am. J. Clin. Pathol.* **6,** 553.

Stein, H., Bank, A., Tolksdorf, G., Lennert, K., Rodt, H., and Gordes, J. (1980). *J. Histochem. Cytochem.* **28,** 746.

Steinbach, M., and Auban, R. (1979). *Arch. Biochem. Biophys.* **134,** 279.

Stenger, R. J., Chabon, A. B., Primus, F. J., and Wolff, W. I. (1979). *Mt. Sinai J. Med. (NY)* **46,** 185.

Sternberger, L. A. (1969). *Mikroskopie* **25,** 346.

Sternberger, L. A. (1979). "Immunochemistry," 2nd ed. Wiley, New York.

Sternberger, L. A., and Joseph, S. A. (1979). *J. Histochem. Cytochem.* **27,** 1424.

Sternberger, L. A., Hardy, P. H., Cuculis, J. J., and Meyer, H. G. (1970). *J. Histochem. Cytochem.* **18,** 315.

Stiller, J. M., and Nielsen, K. H. (1983). *J. Clin. Microbiol.* **17,** 323.

Stollar, V., Harrap, K., Thomas, V., and Sarver, N. (1979). *In* "Arctic and Tropical Arboviruses" (E. Kurstak, ed.), pp. 277–296. Academic Press, New York.

Stoller, R. G., Hande, K. R., Jacobs, S. A., Rosenberg, S. A., and Chabner, B. A. (1977). *N. Engl. J. Med.* **297,** 630.

Sugasawara, R. J., Prato, C. M., and Sippel, J. E. (1984). *J. Clin. Microbiol.* **19,** 230.

Suter, M. (1982). *J. Immunol. Methods* **53,** 103.

Sutherland, S., and Briggs, J. D. (1983). *J. Med. Virol.* **11,** 147.

Sutton, R., Wrigley, C. W., and Baldo, B. A. (1982). *J. Immunol. Methods* **52,** 183.

Svennerholm, A. M., and Holgren, J. (1978). *Curr. Microbiol.* **1,** 19.

Svennerholm, A. M., and Wihlund, F. (1983). *J. Clin. Microbiol.* **17,** 596.

Szumess, W., Much, M. I., Prince, A. M., Hooknagle, J. H., Cherubin, C. E., Harley, E. J., and Bloch, G. H. (1975). *Ann. Int. Med.* **83,** 489.

Takiff, H. E., Strauss, S. E., and Garen, C. F. (1981). *Lancet* **2,** 832.

Tanaka, M., Tanaka, H., and Tshibawa, E. (1984). *J. Histochem. Cytochem.* **32,** 452.

Tandon, A., Zahner, H., and Lammler, F. (1979). *Tropen. Med. Parasitol.* **30,** 189.

Tarkowski, A., Czerkinsky, C., Nilsson, L.-A., Nygren, H., and Ouchterlony, O. (1984). *J. Immunol. Methods* **72,** 451.

Taylor, C. R. (1976). *Eur. J. Cancer* **12,** 61.

Taylor, C. R., and Burns, J. (1974). *J. Clin. Pathol.* **27,** 14.

Thirumoorthi, M. C., and Dajani, A. S. (1979). *J. Clin. Microbiol.* **9,** 28.

Thoen, C. O., Bruner, J. A., Luchsing, D. W., and Pietz, D. E. (1983). *Am. J. Vet. Res.* **44,** 306.

Thompson, R. C. A., and Kumaratilake, L. M. (1982). *Trans. R. Soc. Trop. Med. Hyg.* **76,** 13.

Tiggemann, R., Plattner, H., Rasched, I., Bauerle, P., and Wachter, E. (1981). *J. Histochem. Cytochem.* **29,** 1387.

Tijssen, P. (1985). *In* "Laboratory Techniques in Biochemistry and Molecular Biology" (R. H. Burdon and P. H. Van Knippenberg, eds.). Elsevier, Amsterdam.

Tijssen, P., and Kurstak, E. (1974). *In* "Viral Immunodiagnosis" (E. Kurstak and R. Morisset, eds.), pp. 125–138. Academic Press, New York.

Tijssen, P., and Kurstak, E. (1977). *In* "Comparative Diagnosis of Viral Diseases" (E. Kurstak and C. Kurstak, eds.), Vol. 2, pp. 489–504. Academic Press, New York.

Tijssen, P., and Kurstak, E. (1981). *J. Virol.* **37,** 17.
Tijssen, P., and Kurstak, E. (1984). *Anal. Biochem.* **136,** 451.
Tijssen, P., Su, D.-M., and Kurstak, E. (1982). *Arch. Virol.* **74,** 277.
Tom, H., Kabakoff, D. S., Lin, C. I., Sigh, P., White, M., Westkamper, P., McReynolds, C., and de Porceri-Morton, K. (1979). *Clin. Chem.* **25,** 1144.
Tomagawa, S. (1983). *Nature (London)* **302,** 575.
Towbin, H., and Gordon, J. (1984). *J. Immunol. Methods* **72,** 313.
Towbin, H., Staehelin, T., and Gordon (1979). *Proc. Natl. Acad. Sci. U.S.A.* **76,** 4350.
Towbin, H., Ramjoue, H. P., Kuster, H., Liverani, D., and Gordon, J. (1982). *J. Biol. Chem.* **257,** 12709.
Tracey, D. E., Liu, S. H., and Cebra, J. J. (1976). *Biochemistry* **15,** 624.
Tracey, D. E., Lin, S. H., and Cebra, J. J. (1983). *Biochemistry* **5,** 624.
Troisi, C. L., and Monto, A. S. (1981). *J. Clin. Microbiol.* **14,** 516.
Tubbs, R. R., Gephardt, G., Valenzuela, R., and Deodhar, S. (1980). *J. Clin. Pathol.* **73**(2), 240.
Turner, B. M. (1983). *J. Immunol. Methods* **63,** 1.
Turner, R., Lathey, J. L., Van Voris, L. P., and Belshe, R. B. (1982). *J. Clin. Microbiol.* **15,** 824.
Ukkonen, P., Väisiänen, O., and Penttinen, K. (1980). *J. Clin. Microbiol.* **11,** 319.
Ukkonen, P., Granström, M-L., and Penttinen, K. (1981). *J. Med. Virol.* **8,** 131.
Ullman, E. F., Yoshida, R. A., Blakemore, J. I., Maggio, E., and Ceute, R. (1979). *Biochim. Biophys. Acta* **441,** 235.
Vacca, L. L., Rosario, S. L., Zimmerman, E. A., Tomashefsky, P., Ng, P.-Y., and Hsu, R. C. (1975). *J. Histochem. Cytochem.* **23,** 208.
Van den Hurk, J., and Kurstak, E. (1980). *J. Virol. Methods* **1,** 11.
van der Logt, J. T. M., van Loon, A. M., and van der Veen, J. (1982). *J. Med. Virol.* **10,** 213.
van der Marel, P., Hazendonk, A. G., and van Wezel, A. L. (1981). *Dev. Biol. Stand.* **47,** 101.
Vandesande, F. (1979). *J. Neurosci. Methods* **1,** 3.
van Furth, R. (1975). "Mononuclear Phagocytes in Immunity, Infection and Pathology." Blackwell Scientific Publications, Oxford.
Van Kamp, G. J. (1979). *J. Immunol. Methods* **27,** 302.
Van Knapen, F., van Leusden, J., and Buys, J. (1982). *J. Parasitol.* **68,** 951.
van Loon, A. M., and van der Veen, J. (1980). *J. Clin. Pathol.* **33,** 635.
van Loon, A. M., Heessen, F. W. A., van der Logt, J. T. M., and van der Veen, J. (1981). *J. Clin. Microbiol.* **13,** 416.
van Weemen, B. K. (1985). *J. Virol. Methods* **10,** 371.
van Weemen, B. K., and Schuurs, A. H. W. M. (1971). *FEBS Lett.* **15,** 232.
Varela-Diaz, V. M., Lopez-Lemes, M. H., Prezioso, U., Coltori, E. A., and Yarzabal, L. A. (1975). *Am. J. Trop. Med. Hyg.* **24,** 304.
Vejtorp, M. (1980). *J. Virol. Methods* **1,** 1.
Vervoort, T., Magnus, E., and Van Meirvenne, N. (1978). *Ann. Soc. Belge Med. Trop.* **58,** 177.
Vesikari, T., and Vaheri, A. (1968). *Br. Med. J.* **1,** 221.
Vestergaard, B., and Jensen, O. (1980). *Proc. Intern. Conf. Human Herpes Viruses,* Atlanta, Georgia.
Virelizor, J. L., Allison, A. C., and Schild, G. C. (1979). *Br. Med. Bull.* **35,** 65.

Viscidi, R., Laughon, B. E., Hanvanich, M., Bartlett, J. G., and Yolken, R. H. (1984). *J. Immunol. Methods* **67,** 129.

Voller, A., Draper, C. C., Bidwell, D. E., and Bartlett, A. (1975). *Lancet* **1,** 426.

Voller, A., Bidwell, D. E., Bartlett, A., Fleck, D. G., Perkins, M., and Oladehin, B. (1976). *J. Clin. Pathol.* **29,** 150.

Voller, A., Bidwell, D. E., and Bartlett, A. (1979). "A Guide with Abstracts of Microplate Applications." Dynatech Europe, Guernsey.

von Seefried, A., Bryce, E., Campbell, J. B., Chun, G. D., Laurence, G. D., Scollard, N., and Tsai, K. S. (1981). *Dev. Biol. Stand.* **47,** 91.

Vyas, G. N., and Shulman, N. R. (1970). *Science* **170,** 332.

Wagener, C., Czazar, H., Totovic, V., and Breuer, H. (1978). *Histochemistry* **58,** 1.

Walberg, C. B., (1974). *Clin. Chem.* **20,** 305.

Walbert, C. B., and Won, S. H. (1979). *Ther. Drug M.* **1,** 47.

Walsh, B. J., Wrigley, C. W., and Baldo, B. A. (1984). *J. Immunol. Methods* **66,** 99.

Wang, M. C., Valenzuela, L. A., Murphy, G. P., and Chu, T. M. (1979). *Invest. Urol.* **17,** 159.

Wang, N. S., Shai-Non, H., and Gold, P. (1979). *Cancer* **44,** 937.

Wang, S.-P., Grayston, J. T., Kuo, C.-C., Alexander, R. R., and Holmes, K. K. (1977). *In* "Nongonococcal Urethritis and Related Infections" (E. Hobson and K. K. Holmes, eds.), p. 600. American Society of Microbiology, Washington, D.C.

Webster, R. G., Hinshow, V. S., Laver, W. G. (1982). *Virology* **14,** 93.

Wei, R., and Reibe, S. (1977). *Clin. Chem.* **23,** 1386.

Wei, R., Knight, G. Y., Zimmerman, D. H., and Bond, H. E. (1977). *Clin. Chem.* **23,** 813.

Weidner, N., McDonald, J. M., Treber, V. L., Smith, C. H., Kessler, G., Ladensen, J. H., and Dietzler, D. N. (1979). *Clin. Chem. Acta* **73,** 51.

Weinberger, M. M., and Hendeles, L. (1979). *Postgrad. Med.* **61,** 85.

Weiss, N., Gulzata, M., Wyss, T., and Betschart, B. (1982). *Acta Trop.* **39,** 373.

Welliver, R. C., Kaut, T. N., Putnam, T. I., Sun, M., Riddlesberger, K., and Ogra, P. (1980). *J. Pediatr.* **96,** 808.

WHO (1980). *Bull. WHO* **58,** 585.

Whreghitt, T. G., Tedder, R. S., Nagington, J., and Ferns, R. B. (1984). *J. Med. Virol.* **13,** 361.

Wielaard, F., Denissen, A., Van Elleswijk Berg, J., and Van Gemert, G. (1985). *J. Virol. Methods* **10,** 349.

Wiktor, T. J. (1984). *In* "Control of Virus Diseases" (E. Kurstak and R. G. Marusyk, eds.), pp. 521–528. Dekker, New York.

Wiley, E. L., Murphy, P., Mendelsohn, G., and Eggleston, J. C. (1981). *Am. J. Clin. Pathol.* **76,** 806.

Wilkin, T., Nicholson, S., and Casey, C. (1985). *J. Immunol. Methods* **76,** 185.

Wilson, I. A., Shekel, J. J., and Wiley, D. C. (1981). *Nature (London)* **289,** 366.

Wilson, M. B., and Nakane, P. K. (1978). *In* "Immunofluorescence and Related Techniques" (W. Knapp, H. Holubar and G. Wick, eds.), p. 215. Elsevier North Holland Biomedical Press, Amsterdam.

Wolters, G., Kuijpers, L., Kacaki, J., and Schuurs, A. (1976). *J. Clin. Pathol.* **29,** 873.

Wolters, G., Nelissen, P., and Kuijpers, L. (1985). *J. Virol. Methods* **10,** 299.

Wong, D. T., Welliver, R. C., Riddlesberger, K. R., Sun, M. S., and Ogra, P. L. (1982). *J. Clin. Microbiol.* **16,** 164.

Woodhead, J. S., Kemp, H. A., Nix, A. B. J., Rowlands, R. J., Kemp, K. W., Wilson, D.

W., and Griffiths, K. (1981). *In* "Immunoassays for the 80s" (A. Voller, A. Bartlett, and D. Bidwell, eds.), p. 169. University Park Press, Baltimore.

Wright, J. F., and Hunter, W. M. (1982). *J. Immunol. Methods* **48,** 311.

Wyler, D. J. (1983). *N. Engl. J. Med.* **308,** 875.

Yamamoto, S., and Minamishima, Y. (1982). *J. Clin. Microbiol.* **15**(6), 1128.

Yamaura, N., Makino, M., Walsh, L. J., Bruce, A., and Choe, B.-K. (1985). *J. Immunol. Methods* **84,** 105.

Yolken, R. H. (1982). *Rev. Infect. Dis.* **4,** 35.

Yolken, R. H., and Leister, F. J. (1981). *J. Clin. Microbiol.* **14,** 427.

Yolken, R. H., and Stopa, P. J. (1980). *J. Clin. Microbiol.* **11,** 546.

Yolken, R. H., Kim, H. W., Clem, T., Wyatt, R. G., Kalica, A. R., Chanock, R. M., and Kapikian, A. Z. (1977). *Lancet* **2,** 263.

Yolken, R. H., Wyatt, R. G., Kim, H. W., Kapikian, A. Z., and Chanock, R. M. (1978). *Infect. Immun.* **19,** 540.

Yolken, R. H., Stopa, P. J., and Harris, C. C. (1980). *In* "Manual of Clinical Immunology" (N. Rose and H. Friedman, eds.) p. 692. American Society of Microbiology, Washington, D.C.

Yolken, R. H., Leister, F. J., Whitcomb, L. S., and Santosham, M. (1983). *J. Immunol. Methods* **56,** 319.

Zabriskie, J. B., and Friedman, J. E. (1983). *Adv. Exp. Med. Biol.* **161,** 457.

Zaitsu, K., and Okhura, Y. (1980). *Anal. Biochem.* **109,** 109.

Zentner, B.-S., Margalith, M., Galil, A., Halevy, B., and Sarov, I. (1985). *J. Virol. Methods* **11,** 199.

Zhang Yong-He, Yu Wen-Fang, Tian Zhong-Wen, Chen Qin-Sheng, and Wang Yi-Min. (1984). *J. Virol. Methods* **9,** 45.

Zissis, G., and Lambert, J. P. (1978). *Lancet* **1,** 96.

Zollinger, H. U., and Mihatsch, M. Y. (1978). "Renal Pathology in Biopsy." Springer-Verlag, Berlin and New York.

Zuckerman, A. J. (1981). *J. Infect. Dis.* **143**(2), 301.

zur Hausen, H. (1977). *Curr. Top, Microbiol. Immunol.* **78,** 1.

Zweig, M., Heilman, C. J., Rabin, H., Hopkins, R. F., Neubauer, R. H., and Hampon, B. (1979). *J. Virol.* **32,** 676.

Index

C

D

F

J

K

L

T